Thrive in Ecology and Evolution

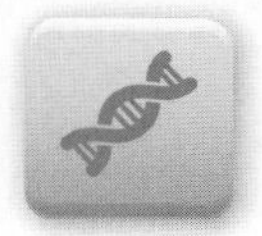

Other titles in the Thrive in Bioscience series

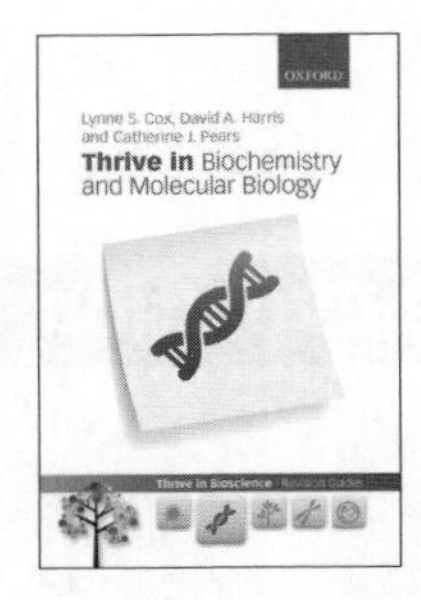

Thrive in Biochemistry and Molecular Biology

Lynne S. Cox, David A. Harris, and Catherine J. Pears

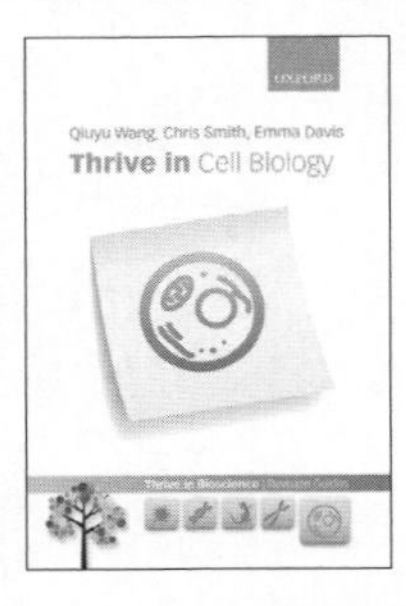

Thrive in Cell Biology

Qiuyu Wang, Chris Smith, and Emma Davis

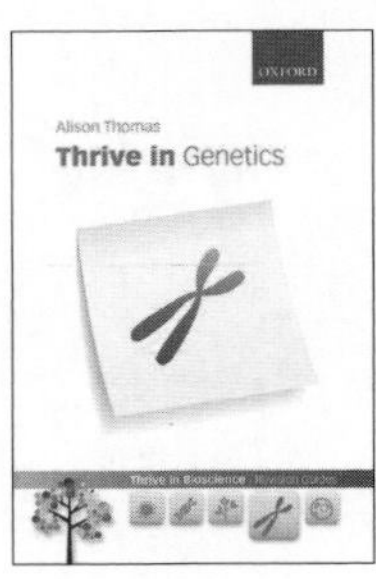

Thrive in Genetics

Alison Thomas

Forthcoming: **Thrive in** Human Physiology

Ian Kay and Gethin Evans

Thrive in Immunology

Anne Cunningham

Thrive in Ecology and Evolution

Alan Beeby
Formerly Reader in Ecology, London South Bank University

Ralph Beeby
Imperial College

Thrive in Ecology and Evolution

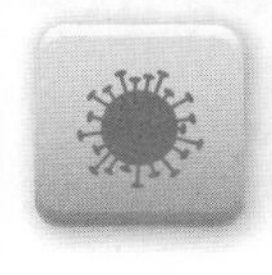
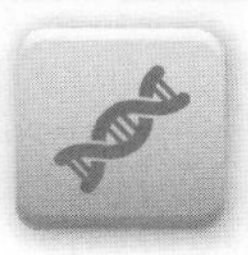

OXFORD
UNIVERSITY PRESS

Great Clarendon Street, Oxford, OX2 6DP,
United Kingdom

Oxford University Press is a department of the University of Oxford.
It furthers the University's objective of excellence in research, scholarship,
and education by publishing worldwide. Oxford is a registered trade mark of
Oxford University Press in the UK and in certain other countries

Impression: 1

British Library Cataloguing in Publication Data
Data available

ISBN 978–0–19–964405–6

Printed in Great Britain by
Ashford Colour Press Ltd, Gosport, Hampshire

Contents

Four steps to exam success

1 *Review the concepts*

This book is designed to help your learning be quick and effective:

- information is set out in bullet points, making it easy to digest,
- clear, uncluttered illustrations illuminate what is said in the text,
- key concept panels summarize the essential learning points.

2 *Check your understanding*

- Try the questions at the end of each chapter and the online multiple-choice questions to reinforce your learning.
- Download the flashcard glossary to master the essential terms and phrases.

3 *Take note of extra advice*

- Look out for revision tips, and hints for getting those precious extra marks in exams.

4 *Go the extra mile*

- Explore the suggestions for further reading listed on the book's website to take your understanding one step further.

Go to the Online Resource Centre for more resources to support your learning, including:

- online quizzes, with feedback,
- a flashcard glossary, to help you master the essential terminology,
- answers to questions in the book, and additional exercises,
- skeletal notes and a conceptual map of each chapter,
- spreadsheets featuring interactive models and calculations.

http://www.oxfordtextbooks.co.uk/orc/thrive/

First words

What you should do

Try to be clear about your objectives in your revision. That is, decide

1. what you need to know.
2. what will indicate that you know it.
3. when you will need to know it.

What you need to know

Ecology and evolution are a set of *ideas* that seek to explain the living world, which you need to understand. Your revision strategy has to distinguish between these concepts and the data and examples which are explained by them.

Ideas need to be understood, facts need to be memorized.

You must understand a concept before you can describe examples to illustrate it. Rote-learning of facts only allows for limited answers. Answers which show deeper understanding achieve higher marks.

Give priority to the concepts early in your revision, accepting that some may need further work to improve your understanding.

In each case you should aim to:

- understand the idea,
- see how it relates to other ideas in ecology and evolution.

It is important that you review and refine your understanding of the major ideas early in your studies: you will almost certainly need to revise and add to your lecture notes. Check your comprehension by writing the concept out in your own words, and then match it against the account here or in the textbooks.

Do not ignore ideas because you have not understood them or because you think you can avoid them in the examination. This is a risky strategy in a subject where all ideas are connected.

Refine your notes as your understanding improves and apply the concepts to case studies you cover in lectures and in your reading. These studies should prompt questions in your mind. Add these to the summaries of the example and revise them along with the detail.

Ensure your revision notes develop and reflect the work you do as the course progresses.

What will indicate that you know it

You understand an idea if you can write a short account showing how it explains a set of observations. Check that you and the textbooks agree on the basic principle and that you can write an accurate summary of it.

Be aware that there may be several interpretations or explanations for a topic. You will demonstrate greater understanding if you can set out both sides of an argument with supporting data and examples. An excellent answer will critically evaluate an idea, or suggest further observations that might decide between alternative explanations.

You can be confident that you understand a concept when you can apply it in different contexts. This has three important implications for your revision process:

1. you will score highly if you can use the concept outside the context in which it was taught;
2. it will inform your revision in other areas and allow you to give fuller answers;
3. it will stop you thinking in small boxes. You will see the connections between subjects, and demonstrate in-depth understanding.

This is also a good reason for using the primary scientific literature to find different case studies or examples to illustrate a principle, rather than just relying on those in the standard textbooks. Equally, try to draw upon more than one textbook.

This guide is designed to facilitate understanding: it concentrates on ideas and the connections between them. It does not provide examples or data.

Throughout the book we highlight the connections between topic areas to help you build a mental map as the lecture programme proceeds. A conceptual map for each chapter is provided on the book's website. You can use this to identify gaps in your understanding and where you need supporting data.

When you will need to know it

(a) Take control of your revision

Begin your planning early…well in advance of the examination period, preferably starting with the teaching programme.

You need to be *proactive* and explore the ideas: do not let the ideas wash over you and hope that you will absorb them by osmosis. It is very difficult to gain a full understanding *passively*, by memorizing a set of lecture notes. Instead, use this guide to:

- anticipate the concepts before the lectures,
- think about the concepts and raise questions in tutorials soon after the topic has been taught.

Give yourself time to understand: do not hope that this will fall into place the night before the examination.

Ensure that you have a good set of notes, organized for easy revision: you can use the facility on the website to produce a fair set of notes (see the section on the website), structured by the main ideas:

- in creating the notes, add a commentary or critique, supported by examples. Include an evaluation of the data or methodology of a case study and suggestions for further experiments. This is another indication that you have a good understanding and will demonstrate your reading around a subject;
- allow the revision notes to develop as your course progresses: regard them as a work in progress and revisit them frequently;
- aim to finish with a set of notes in which you have confidence, comfortable with the ideas.

This will make the examinations and the revision process less fraught.

In some cases there may be numerical examples that need to be worked through. Use the exercises at the end of each chapter and on the website to test these skills. Ensure that you are confident in these methods, especially if they are likely to feature in examinations.

(b) Schedule your revision

Use the lecture programme to plan a schedule of revision. Begin by extracting the main ideas and noting their connections with each other. Then consider related units and see if there are obvious areas of overlap.

Try to construct a conceptual map of the topics or use those on the website: many people find these useful for quick revision. Add flags to relevant studies or authors.

Towards the end of the unit, create your revision schedule. In particular, identify:

- concepts which need work, either to improve your understanding or to review alternative arguments;
- missing notes, especially the need for examples;
- the connections between ideas, especially the 'high currency' concepts that support more than one topic area;
- the sequence in which these concepts should be revised (do not assume this is the sequence in which they have been delivered).

Now apply deadlines: dates when your work on the main topics needs to be completed and your understanding in place.

The result should be a schedule over an extended period, completed well in advance of the examination, and before your final review.

A word about ideas and this guide

This book straddles two subjects but because ecology and evolution share so many ideas they can be seen as two sides of the same coin. An organism is fitted to its habitat by the selective pressures in its environment. Darwin's theory is ultimately ecological.

The properties of species, populations, and communities are shaped by selective pressures operating on individual organisms. An ecosystem is the summation of these interactions and the biosphere the summation of these ecosystems. This requires an understanding of processes from the level of the gene to the natural history of the planet. Our task in this guide is to make the connections as we move up these levels.

We do this by using ideas that have explanatory power at most levels in this hierarchy. Evolution by natural selection is the most fundamental theory for the whole of biology, but we need to approach it from an ecological perspective.

Ecological niche

This guide uses ecological niche as its unifying concept, viewing an individual or a species as the product of past selective pressures.

The *ecological niche* of a species is the position it occupies in relation to its biotic and abiotic environment. It is sometimes described as a species 'role', emphasizing its interactions with other members of the community and function in the ecosystem. It is the totality of the selective pressures that determine a species' fitness or reproductive success.

It is important to see why the concept of niche is unifying:

- it is the ecological complement of the species' concept: a niche is the ecological space to which a species has been fitted by natural selection;
- these inherited adaptations distinguish one species from another and by which they are identified;
- it helps describe a species' significance for other species or ecological processes;
- it recognizes that species change as their environment changes.

We say more about this in Chapters 1 and 2. Chapter 1 provides an overview of the main themes through the guide, to show how they are developed in subsequent chapters. This should be a useful (and quick) check on your understanding before and during the revision process.

Remember that ecology deals principally with the higher levels of biological organization: from the organism to populations, communities, and ecosystems. We therefore have to assume you understand some key science below the level of the organism, especially in genetics and biochemistry. This assumed knowledge is stated at the beginning of each chapter.

In addition, the guide:

- summarizes the key concepts at the beginning of each chapter;
- posts flags to related concepts elsewhere in the book;
- provides questions that test understanding, both at the end of each chapter and on the website;
- provides a glossary. Terms highlighted in **bold** are defined in the glossary.

The guide does not:

- provide data to illustrate the concepts;
- represent all the material you will need for revision: this differs between courses and units, and you need to customize your revision notes to reflect your lecture programme and your reading around the subject.

The guide does provide a basic account of the main concepts, a scaffold on which you should hang examples relevant to your particular course.

In summary

Be proactive: begin, if you can, when the lecture programme starts. Address weaknesses as they arise.

Decide what you need to know. Create a schedule and work to deadlines.

Produce a set of comprehensive and trusted revision notes.

Ensure you have illustrative examples for the main concepts before you begin your revision.

Ensure you revision is *active* and not *passive*. Create conceptual maps, add notes, review examples or experiments, write out summaries, and so on.

Decide how well you understand the main concepts based on your capacity to:

- give a clear and precise account of their argument;
- offer an alternative argument;
- apply them in different contexts;
- evaluate the significance of data that test the argument;
- demonstrate how your examples illustrate the concept;
- make the connections with other concepts;
- answer the questions here and on the website.

Aim to become *comfortable* with the main ideas in ecology and evolution; to play around with them, to see where they might apply, to review and explore their value in different contexts.

The web pages

http://www.oxfordtextbooks.co.uk/orc/thrive/

There are several resources in the Online Resource Centre that accompanies this revision guide to help you:

1. A conceptual map provides an overview for each chapter. You will see the conceptual maps flagged at the beginning of the chapter.
2. Skeletal notes for each chapter. These provide the main concepts and the links between ideas in other chapters. Download these to add additional notes, or to re-format your lecture notes. You can also download figures from the book to add to your notes.
3. Models or calculations for most of those covered in the text. Use these to explore the principle and the properties of the models. You will see these models and calculations flagged at the relevant points in the text.
4. Answers to the questions at the end of each chapter and additional exercises with answers.

All exercises, both in the book and on the website, are designed to test understanding rather than recall. From these you can judge how your understanding is developing.

We especially encourage you to attempt the questions on the website. The model answers give an indication of what examiners will look for, but also develop the concepts being considered. You should find these helpful if you need further explanation.

Further questions and answers are available in our textbook *First Ecology* (Beeby and Brennan, 3rd edition, Oxford University Press, 2008). This also provides examples and case studies for some of the concepts covered here. Its Online Resource Centre provides hyperlinks to the literature cited in that book, giving the reader access to the original research papers. This site also has a virtual field course, using video footage of various ecosystems, and spreadsheets of the data collected. Users can analyse this data in structured exercises, supporting several topics covered here, including succession, species–area relations, and so on.

Alan Beeby
Ralph Beeby

London
September 2012

Acknowledgements

We are grateful for the excellent editorial support of Jonathan Crowe (once again), Alice Mumford, and Angela Butterworth at Oxford University Press, and freelance copy-editor Nik Prowse, and proofreader Emma Tuck. We thank our four anonymous reviewers for their insightful comments and useful recommendations.

We also thank Jackie and Kate Beeby, who have played various roles in the gestation and delivery of the whole project.

1 Fundamentals of Ecology

Key concepts

- The fundamental unit in biology is the species. The 'biological' species is a group of individuals sharing most of their genotype, able to interbreed and produce viable and fertile offspring.
- In taxonomy the binomial system gives all multicellular species a unique combination of generic and specific names.
- Systematics is the organization of species into a nested hierarchy of groups or taxa, progressively divided into smaller taxa of increasing similarity, reflecting their phylogeny or evolutionary relationships.
- The genome of a species is primarily the result of natural selection and its adaptations to a particular habitat and way of life: its ecological niche.
- Adaptations improve the fitness of an individual, measured by the transmission of its genetic information to the next generation.
- Responding to different selective pressures, an organism will evolve adaptations that may incur costs as well as conferring benefits. Natural selection will, over generations, find the balance of costs and benefits that maximizes reproductive success.
- Both biotic and abiotic factors create gradients that structure ecosystems and communities, stratifying their space and generating periodicities. Species have to adapt to this ecological space and time.
- Populations, communities, and ecosystems are structured by these gradients.

Assumed knowledge

The coding of genetic information in multicellular organisms and how this is passed on by sexual reproduction.

Note that this chapter provides an overview of the main concepts discussed in the rest of the book.

A conceptual map for this chapter is available on the book's website: go to http://www.oxfordtextbooks.co.uk/orc/thrive/ or scan this image:

1.1 THE SPECIES

Organisms were first grouped by their appearance and early classifications were based on similarities and differences in their morphology. Eventually, these **morphological species** were based on 'type specimens' (and consequently, are sometimes referred to as typological species). This remains the basis of the classification of most higher plants and animals.

Naming and classifying organisms

- Classification—grouping entities by shared traits—was one of the earliest forms of science. Chemistry advanced rapidly after the advent of the periodic table and the grouping of elements by shared properties. Classifications prompt questions about the relationships between groups.
- A fundamental property of life is the capacity of a group of organisms to breed with each other. The collection of all interbreeding populations represents a species.
- Species can themselves be grouped to create a higher-level identity. Repeating this process with successive groupings produces a nested hierarchy: more inclusive groups occur at the higher levels.
- This is possible because of the shared ancestry of all organisms. A classification aims to create a hierarchy which maps the evolutionary history or **phylogeny** of its species (Table 1.1).
- **Systematics** is the set of principles used to locate species in this system and by which the hierarchy is constructed.
- **Taxonomy** is the conventions used to name species and the higher-level groups.
- Species share much of their genetic history and many of their traits with other members of their genus; at the higher levels they have less in common. Today we

Revision tip

Ensure that you know the difference between taxonomy, systematics, and phylogeny.

Taxon	*Horse*	*Zebra*	*Giraffe*
Kingdom	Animalia	Animalia	Animalia
Phylum	Chordata	Chordata	Chordata
Class	Mammalia	Mammalia	Mammalia
Order	Perissodactyla	Perissodactyla	Artiodactyla
Family	Equidae	Equidae	Giraffidae
Genus	*Equus*	*Equus*	*Giraffa*
Species	*ferus*	*grevyi*	*camelopardalis*

Table 1.1 The classification of three large mammals.

The giraffe belongs to the Artiodactyla, the even-toed ungulates; the Perissodactyla are the odd-toed ungulates.

Horses, asses, and zebras belong to the same genus. All horses belong to the species *Equus ferus*, but there are two extant (living) subspecies: the wild horse *E. f. przewalskii* and the domesticated horse *E. f. caballus*.

The genus *Equus* has subgenera for both the asses and the zebras. For zebras these are *Hippotigris* (which includes the mountain zebra *Equus zebra* and the plains zebra *Equus quagga*) and *Dolichohippus* (which only includes *Equus grevyi*, found in the more arid regions of northern East Africa). Grevy's zebra is able to produce fertile hybrids with the plains zebra.

Giraffes are divided into nine subspecies, principally on the colour and patterns on the coat. Each is confined to relatively well-defined regions throughout sub-Saharan Africa and some are thought to be close to being distinct species; perhaps as many as six are reproductively isolated from each other, based on their genetics.

can determine the phylogeny, and often the age, of a group by comparing the genotypes of its extant (living) species. Many viral and prokaryotic species are identified by their **genome**.

The binomial system

- The original hierarchical system of Linnaeus was based on seven levels:
  - Kingdom,
  - Phylum (Division),
  - Class,
  - Order,
  - Family,
  - Genus,
  - Species.
- These groups are called **taxa** (singular: **taxon**). Here the most inclusive taxon is the kingdom, encompassing all the lower taxa.
- Above the species, the lowest level is the genus. Species belonging to the same genus are termed 'congeneric'.

Revision tip

You need to know these seven taxa in the correct order.

- The binomial system is a nested hierarchy because lower taxa are grouped in the taxon above. For example, the fossil record suggests we are the only extant species of four (at least) in the genus *Homo*; *Homo* is one of four extant genera in the family Hominidae.
- With the exception of the species, each taxon is defined by its relative position. For this reason a taxon in one hierarchy may not be equivalent to its counterpart in others.
- Phyla only apply to animals, and are replaced by Divisions when classifying plants and bacteria. Viruses are classified principally into families and genera.

Revision tip

Identify the most used taxon for each major group of organisms, especially those you have studied on field courses or in laboratory work. For example:

in the chordates it is the class (mammals, bony fish, reptiles, etc.);

for insects it is the order (beetles, flies, bugs, etc.);

for flowering plants it is the family (umbellifers, grasses, composites, etc.).

You should know these for groups that feature often in your course.

- Modern classifications subdivide these seven categories: there may be up to 25 levels for an animal, whereas some plants may be divided into four taxa *below* the level of species.
- These additional levels attempt to create a natural classification, to reflect the phylogeny of a species. Evolution produces gradual change on which we impose our taxa and these demarcations can be highly contentious.
- Only traits that are inherited can be used to infer the phylogeny of a group.
- Many groups, especially the invertebrates and microorganisms, are poorly described. With the advent of **genomic analysis** all phylogenies are under review.
- These techniques are refining (often confirming) the systematics of groups originally based on typological species. This has led to some major revisions and surprise connections between groups.

Revision tip

You should have one example of how genomic analysis has revised the systematics of a group of plants or animals.

See Box 1.1 for an outline of genus and species naming conventions.

The species as the fundamental unit in biology

The **species** is the only taxon that is functionally defined and it is thus the fundamental unit in multicellular biology: it demarcates a genetically coherent group in which evolution can occur over generations. The species is a collection of

Box 1.1 Key technique

Cite scientific names properly to avoid losing marks

A species name is written in italics (or underlined if hand-written): *Homo sapiens*

The generic name begins with a capital letter: *Homo*

and the species names with a lower-case letter: *sapiens*

Notice that the combination of the two names is unique: generic names are shared by different species belonging to that genus: *Homo sapiens, Homo habilis.*

The specific name may be used in different genera: *Potentilla palustris* and *Rumex palustris.* However, the combination of the two names is unique: there is only one *Homo habilis.*

The name is given in full on first citation. In scientific papers the first author's name follows immediately afterwards: *Homo sapiens* L. ('L.' is the standard abbreviation for Linnaeus who first classified this species).

If the author's name appears in parentheses the species was originally placed in a different genus: *Cantareus aspersus* (Müller), the common brown garden snail.

A subspecies adds a third name, also with a lower-case letter: *Cantareus aspersus maxima.*

After first citation, this name can be abbreviated: *C. aspersus* or *C. a. maxima.*

If there is the possibility of confusion with another genus, the abbreviation has to distinguish between genera: the clam *Chlamys varia* would thereafter be abbreviated to *Ch. varia* to avoid confusion between *Cantareus* and *Chlamys.*

Where a species is not specified for a genus, the abbreviation 'sp.' is used, or with several congeneric species 'spp.': *Chlamys spp.*

The names of taxa are often (but not always) descriptive, especially at the generic and specific level, requiring some familiarity with Latin or Greek. This description may refer to the organism (for example, *hirsuta* would suggest the species is hairy) or its habitat (*palustris* implies a swamp or marsh).

individuals that could successfully breed together. Ernst Mayr developed the **biological species concept**, and defined it as 'a reproductive community of populations (reproductively isolated from others) that occupies a specific niche in nature'.

The biological species concept demarcates:

- individuals with a common ancestry, descended from an ancestral population;
- the smallest genetically distinct group of organisms that share this descent;
- individuals able to mate and swap genes with each other (or which are self-fertile) to produce viable offspring (able to become adult), and are themselves fertile.

The species is said to share a common **gene pool**.

➔ *Section 3.1, Types of species*

- Among organisms that reproduce sexually, both partners have originated from the same gene pool and thus have reproductively compatible genotypes.

- An individual transmits its genes to the next generation when it reproduces. More precisely, it is the instructions coded in the genes carried by the gametes—*information*—that is inherited.

Time and the definition of a species

- Species change over time and, in the process, become genetically distinct from their ancestors.
- The **evolutionary species concept** (see Box 1.2) recognizes that any demarcation between species is contingent upon time: a species is a discrete evolutionary lineage with the capacity to change in the future, just as it has in the past.
- One type specimen represents a single phenotype and gives no indication of the morphological or genetic variation in the population from which it was collected.
- Similarly, the individuals we examine today may also have significant genetic differences from the type specimen collected 100 years ago.
- There are fossils that appear identical to the living specimens we study today; the latter are known as 'living fossils'. These lineages can only be regarded as morphological species (e.g. the coelacanth (*Latimeria*) and the dawn redwood (*Metaseqouia*)) since their genetic signature has almost certainly changed.

Box 1.2 — ***Looking for extra marks?***

The evolutionary species concept

Consider the various ways in which our functional definition of a species would have to change if, rather than relying on its current genetic identity, we could include a species' genetic history. For example, genomic analysis of museum specimens might find that the type specimens of plants from 100 years ago are genetically distinct from their namesakes growing today at the original site of collection. Should they be assigned to a different subspecies?

Until a few years ago, we believed we shared little of our genetic history with the Neanderthals, yet recent genomic studies suggest Neanderthals bred with modern humans when we arrived in Europe and Asia 40 000 years ago. Some modern human populations carry Neanderthal genes, so at what level are we separated from them? Should we regard them as a distinct species (*Homo neanderthalensis*, as they appear in many textbooks) or a subspecies of our species (*Homo sapiens neanderthalensis*), just as we are classified (*H. s. sapiens*)?

Consider the recent reports of crosses between polar bears and grizzly bears in the sub-Arctic of North America. Although they are clearly adapted to different niches, do 'grolar bears' indicate that the gene pools of the two bears were only ever separated geographically? Were these species only morphologically distinct as long as they never met? Does climate change mean they are about to share the same gene pool?

Think of the questions raised about the biological species concept and the circumstances when the evolutionary species concept might be more useful.

1.2 ECOLOGICAL NICHE

An organism is fitted to a habitat by natural selection; that is, its genetic inheritance codes for traits that promote its survival and reproduction. A polar bear is adapted to life on the ice sheet of the Arctic and the grizzly bear to the forests and tundra of sub-Arctic North America. Their coat colour is their most obvious adaptation, the phenotypic expression of the genes they have inherited. These different coat colours can be explained by:

- **proximate** explanations: *how* the difference is produced (the details of the structure and pigmentation of the hairs);
- **ultimate** explanations: *why* different coat colours are adaptive.

Proximate answers describe the detail of how a phenotype is expressed; ultimate answers suggest why the code for that trait is found in an individual. Most of the discussions in the rest of this book concern the ultimate explanations for adaptations.

Defining niche

- A multitude of factors act as selective pressures and determine where a species can survive and reproduce. Together they define the **ecological space** it occupies. This position, along with its response to these pressures, is the natural history or **autecology** of a species.

  ➔ *Section 4.1, Autecology*
- We distinguish between the physical or **abiotic** factors to which a species is adapted (moisture, temperature, salinity, season, time of day, etc.) and the **biotic** factors (the other species with which it interacts).
- For each environmental factor there will be a range within which a species can survive. All these factors and their ranges, together, describe its **ecological niche**: where and how it lives.
- The niche is the combination of all the selective pressures to which a species has evolved adaptations (Figure 1.1). The genes that code for these traits distinguish it as a species and give it a distinct gene pool.
- The niche can thus be seen as the complement of the biological species concept, the ecological space to which a species is fitted.
- Because a niche will change through time it is also the complement of the evolutionary species concept.

  ➔ *Section 2.1, Tolerance limits*

Ecological roles or function

- Selective pressures will cause the loss of the poorly adapted. The better adapted flourish because they have fewer checks on their performance, produce a larger number of offspring, and more of their genes survive into the next generation.
- An organism must acquire the resources to grow and reproduce if it is to pass on the code written in its genes. But it will also represent a resource to other species.

Ecological niche

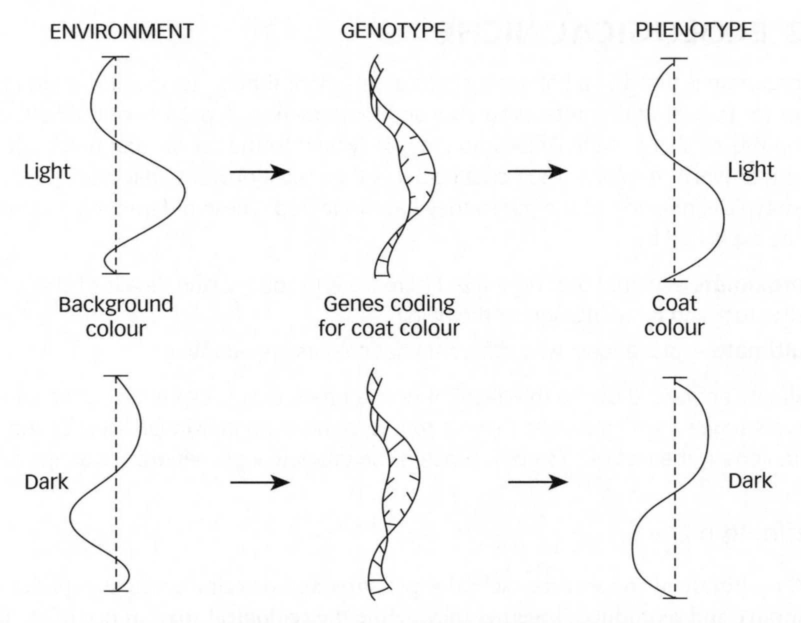

Figure 1.1 A schematic representation of adaptation. This example depicts the match between hypothetical 'spectra' for the background and coat colour, adapting polar bears (*Ursus maritimus*) and grizzly bears (*Ursus arctos horribilis*) to their habitat.

Against a light-coloured background genes coding for a light coat will favoured, and a white bear will be a more successful predator. A bear with a dark coat will best be fitted to a predator's niche in a forested habitat. We assume, of course, that both bears benefit from being camouflaged.

- A species can be classified by the resources its exploits and by the resource it represents; so, within a community of producers, consumers and decomposers, herbivores and carnivores, and hosts and parasites, a species can play several roles.
- Just as a species' abiotic environment represents a complex ecological space, so does its biotic environment; that is, its interactions with other species and the variety of roles it plays in the community.
- Its role is most easily quantified by the transfer of energy or nutrients it facilitates through the ecosystem. However, it might also be defined by its other ecological functions: perhaps as a decomposer, a host for parasites, or a refuge from predators.

  ➔ *Section 2.2, Niche*
- Note that natural selection does not favour adaptations that allow a species to fulfil a role for the *good of the community*: this would imply purpose or design. Natural selection does not work to a design. A genotype is favoured only if it improves that individual's **fitness**.
- For example, if fixing more nitrogen allows a bacterium to leave more offspring, then selection will favour the more effective nitrogen-fixers, *not* because the plant community benefits from a nitrogen-rich soil.

  ➔ *Section 2.3, The principles of natural selection*

- Adapting to one pressure may impair an organism's response to another factor. Over generations, with individuals having different balances between these factors, natural selection will find the **trade-off** representing the best compromise. Again, the genetic code of those with the most successful trade-offs will dominate the gene pool.
- In some cases, greater reproductive success is realized by the offspring rather than the parent. However, the outcome is the same: their genotype is more common in the gene pool.

➔ *Section 4.2, Trade-offs and fitness; Section 2.5, Inclusive fitness*

Revision tips

Whereas phenotypes succeed or fail, remember it is the genetic code alone that is inherited.

Populations and communities do not inherit identities other than the information written in the genes of individuals.

Ecological space and time

- Many of the key factors exist as a gradient, where the factor's value or intensity changes over space; for example, the moist soils next to a river become drier further away from the river.
- Similarly, gradients exist through time, such as the change in light intensity with the time of day or time of year.
- A species will be found somewhere along a gradient, within the limits to which it is adapted: the distribution of a plant, for example, may reflect average moisture levels in the soil, or the light intensity at different times of the year.
- An organism's reproductive output will be maximized where conditions are close to those to which it best adapted (Figure 1.2). This prime position, over all key gradients, is termed its **fundamental niche**.
- Many individuals may not be able to occupy their optimum on all gradients: this space may be occupied by other members of the population or by other species. They are then constrained to a marginal habitat, an ecological space inferior to this optimum: its **realized niche**.

➔ *Section 2.2, Intraspecific competition and niche*

- Ecosystems have periodicities to which its species have to adapt: diurnal, tidal, and seasonal cycles. **Ecological time** governs the availability of resources, the composition of the community and the structure of the ecosystem.
- Members of these communities, visitors and residents, have to adapt their feeding and reproductive strategies to these temporal gradients.
- Gradients in space and time interact and may not be independent of each other. A species may not be able to occupy the optimum position for one factor because another factor prevents it from doing so. For example, the optimum nitrogen

Ecological niche

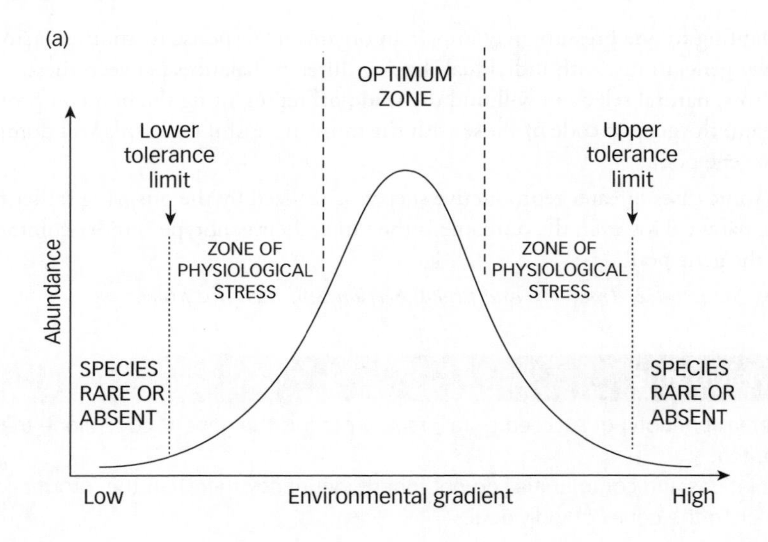

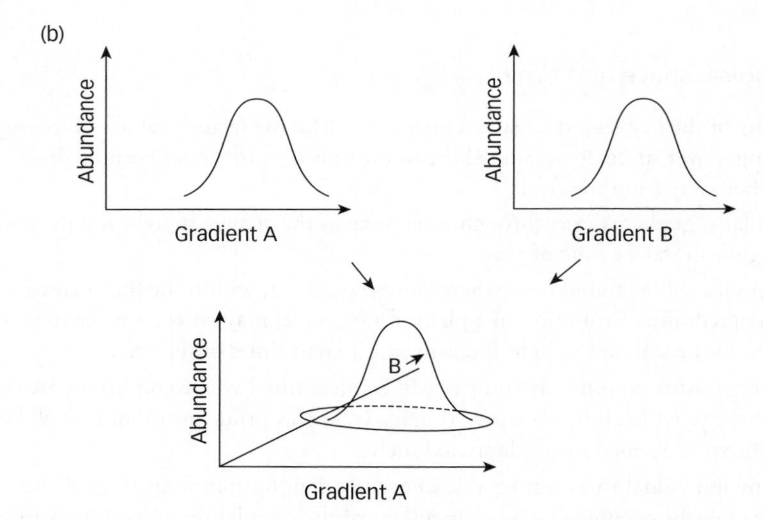

Figure 1.2 (a) The limits of tolerance and the optimum range for a species along an environmental gradient, such as temperature, salinity, or time of day. Some measures of a species' performance (such as its growth rate, reproductive rate, or abundance) can indicate its optimum range for that factor. Performance declines outside this range, when more resources are devoted to adapting to the poor conditions. Beyond its tolerance limits an individual cannot adjust its physiology or behaviour and is unable to maintain itself over the long term.

(b) Species have to adapt to more than one gradient. We can represent the range for two factors together as a three-dimensional volume: the combination of the two tolerance curves creates a bell-shaped 'space'. A species is adapted to a large number of environmental factors, abiotic and biotic, which can interact in complex ways, and creates a multidimensional ecological space which is not easily represented on paper.

levels for a plant may be found in soils too cold for it to survive all year round. It makes a trade-off between two key factors.

- In its optimum ecological space an individual will maximize its reproductive output. Here more resources can be devoted to reproduction. Elsewhere resources have to be directed towards survival, accommodating the poorer conditions.
  ➔ *Section 2.1, The costs of acclimation*
- Gradients change: over days, over seasons, and over the longer term. These temporal gradients may allow two species exploiting the same resource to co-exist, preventing one out-competing the other.
- Species change as gradients change. Individuals alter their behaviour, physiology, and growth with the season. Consistent change over the longer term will cause populations to change their genotype through natural selection.
  ➔ *Box 5.3, Niche differentiation*

Cost-benefit analysis

- Any organism has to balance the benefits of acquiring a resource against the costs of its acquisition.
- A predator must balance the energy costs of catching a prey against the energy gained from consuming it. If the costs outweigh the benefits, the predator has less energy to devote to reproduction, reducing its fitness.
- Over generations, using the variation between individuals, natural selection selects between alternative strategies of balancing the costs and benefits.
- Natural selection does not always produce the best solution: any adaptation will be constrained by existing adaptations and the variation available in the population.
- Individuals able to achieve a favourable balance, across different environmental factors, will maximize their fitness and enjoy greater reproductive success.
- A **cost-benefit analysis**, assessing the costs and advantages of a behaviour or adaptation, can explain aspects of a species' autecology.
- For example, it can explain different reproductive strategies: whether an individual should invest its resources in a single reproductive event, producing many offspring, or several events, each producing a small number of offspring.
- In this case, a key factor is the predictability of the habitat: plants or animals adapted to an unpredictable environment will tend to reproduce at the first opportunity and then die. This strategy requires most resources to be devoted to reproduction.
  ➔ *Section 4.2, Life history and reproductive strategies*
- A large number of offspring from a single reproductive event improves the chances that some (occasionally many) survive to adulthood. In an unpredictable environment most will not.
- A predictable environment favours repeated reproduction. The species that use this strategy spread their risk over time: their costs are low because few offspring are produced, but this can be repeated several times.

- The success of either strategy depends on the availability of resources or, more generally, how closely the adaptations of the offspring match the habitat in which they find themselves.

  ➔ *Section 4.2, Reproductive strategies*
- **Optimality theory** suggests that natural selection will tend to optimize any trait to maximize reproductive success.
- It compares allocation strategies and tests whether selection has arrived at one close to the optimum. Because of the advantage they confer, such strategies are not easily replaced.
- Such analyses can extend across all aspects of the organism's life history. The ecologist has to decide which variables are most important for each species.
- Most especially, we need to measure the traits subject to selection, although the selective advantage may not always be obvious. Delayed reproduction is one example.

  ➔ *Box 4.1, Trade-offs in life history strategies*
- Species may not operate at the optimum predicted by the theory because they make trade-offs between different factors, or because they are still in the process of adapting to their ecological space.
- For this reason, few species are perfectly adapted to their niche, and some, like the burrowing parrot or the tree-kangaroos, can survive only because of the absence of significant competitive pressures from other species.

Niche, competition, and speciation

Living in marginal habitats will mean fewer offspring and fewer genes passing to the next generation. Those close to their optimum conditions will enjoy greater fitness.

- Selective pressures may thus be more intense on individuals in marginal habitats, living in an ecological space to which they are poorly adapted.
- Under selective pressure, those able to use resources efficiently, and produce many viable offspring, are favoured over the rest. By adapting to the conditions, some may come to occupy a greater range of this ecological space.
- If, by its activity or numbers, a population or species can dominate a resource it will reduce the availability to its competitors. This increases the pressure on other species and individuals.
- A population closely adapted to localized conditions will have a distinct genotype, especially if there is limited movement and genetic exchange with its neighbours. Such local races are called **ecotypes**.
- Ecotypes are common in sedentary species, especially among plants. One example is the tolerance of some grass species to toxic metals in mine spoil heaps. The lower fitness of hybrids between the local and neighbouring populations helps to preserve the genetic identity of the ecotypes.
- Ecotypes may be in the process of evolving into a distinct species. With an increasingly distinct gene pool they follow a different 'evolutionary trajectory' from the original population. Such locally differentiated populations have been termed **evolutionarily significant units** (ESUs), recognizing that they represent

incipient species. Some of our conservation efforts are aimed at safeguarding ESUs from extinction.

➔ *Box 2.2, Ecotypes*

1.3 POPULATIONS, COMMUNITIES, AND ECOSYSTEMS

Regarding niche as the complement of the biological species helps us to link evolutionary processes at the species and population levels with the ecological processes found in communities and ecosystems.

Populations

- Populations close to their fundamental niche have fewer checks on their growth. Individuals here are likely to have the highest fitness, have a higher survival rate, and produce most offspring.
- Population growth and survival will be lower where selective pressures are intense, where the conditions are not close to those to which the species is adapted.
- The rate of population growth can thus be linked to the fitness of its individuals and the availability of resources.

  ➔ *Section 4.4, Population growth in a limited environment*

  The gene pool of a species will change faster when a population is poorly adapted to its habitat–as long as there is variation between individuals.

- A changing habitat generates different selective pressures. A persistent change will push the population towards a different set of traits. Over several generations this directional change will be observed in both the genotype and phenotypes of the population.
- Without significant migration or a high mutation rate in the relevant genes a population with little genetic variation will change slowly.
- Competing species, predators, or pathogens will all slow a population's growth. However, these interactions can themselves evolve over time, and thereby continually re-shape the ecological space of a niche.
- Few populations or species are removed from selective pressures for any length of time. The rare examples of 'living fossils' invariably come from long-standing and constant habitats, such as the deep waters of oceans.

Make the connection

Note here the capacity of adaptation and niche to explain the population ecology and the age of some taxa. As ultimate explanations they have explanatory power over a wide range of biological phenomena. This power is demonstrated again under the headings Communities and Ecosystems.

➔ *Section 1.3, Communities; Section 1.3, Ecosystems*

Communities

- Communities are structured by the interactions between their species and the abiotic conditions in which they become established.
- Species can consume or be consumed, cooperate, or compete with each other. As part of the biotic environment these associations exert selective pressures, and again each individual will adapt to maximize its fitness: by escaping consumption, or by consuming, cooperating, or competing more efficiently.

  ➔ *Section 5.1, Species interactions*
- Associations which benefit both species develop from these adaptations and increase the benefits and reduce the costs to each partner. They will persist as long as there is a positive net balance for both species.
- Competition for a limited resource, either between individuals of the same species, or between species, will select the most efficient individuals. Those able to secure the most resource at least cost will have the higher fitness.
- A change in predator numbers or behaviour can demand rapid change in its prey, both in its phenotype (such as changes in behaviour) or in its genotype over generations. In the same way, change in the prey can demand change in its predator.
- This is **co-evolution**, where a change in one species prompts adaptive change in another. Both species evolve as they adapt to changes in the other. This occurs in all forms of association between species.
- In this way, a community of species can become tightly integrated, developing very close associations.

  ➔ *Section 5.5, Co-evolution; Section 6.3, Community integration*
- In many established communities, some species are only able to occupy a niche because of the conditions or resources provided by other species.
- Such close associations are one reason why communities develop through time in a succession. Ecologists debate the extent to which such associations determine the nature of ecological communities.
- They may also contribute to their stability and this has been linked to the high species diversity found in some communities.
- High species diversity is itself an indication of finely divided niches and the complexity added to the ecological space by biotic factors.

Ecosystems

- An **ecosystem** consists of an ecological community set in its abiotic environment.
- Which communities develop in a location depends principally on the abiotic conditions. In terrestrial ecosystems the key factors are the average temperature and the availability of water.
- Climatic regions tend to have distinct terrestrial plant communities, or **biomes**. These broadly follow lines of latitude across the planet, although local conditions can disrupt this simple pattern.

➔ *Section 8.3, The major terrestrial biomes*

- Species in a biome are adapted to the prevailing abiotic conditions: the seasons, availability of minerals, levels of primary production, and so on. These create the gradients that define the ecological space and time of these ecosystems.
- Many ecosystems are stratified into distinct zones according to the abundance of resources or the nature of their physical environment. This stratification is found in both terrestrial and aquatic ecosystems.
- Stratification serves to partition an ecosystem, creating different ecological spaces and niches, and consequently the diversity of species.
- This partitioning helps us to model these systems. We can use models to quantify and compare ecosystems, identify shared features, and understand their organization and regulation.

➔ *Section 7.1, Systems ecology*

- By following the acquisition and transfer of resources (energy, major nutrients, or even pollutants) from the physical environment, through the various components of the ecosystem we can measure:
  - i. a budget for the entire ecosystem and the flux of those resources through the ecosystem;
  - ii. the efficiency of these transfers and its effect on population sizes;
  - iii. the role of individual species in these processes;
  - iv. the impact of the loss of key species on ecosystem function.
- This systems approach allows us to compare ecosystems, rather as we might compare the performance of machines. We can also look for general rules by which they may be organized:
  - i. to find any controlling factors common to very different species assemblages;
  - ii. to understand why performance should differ between very similar ecosystems;
  - iii. through experimentation, to establish which niches are important to performance, and the capacity to withstand disturbance.

This method requires a description of the key functional roles—the niches—the species occupying them, and how they respond to change. Such knowledge is essential for the conservation of habitats.

Looking for extra marks?

Ecology can itself be grouped into a nested hierarchy, its levels of organization:

individuals,
populations,
species,
communities,
ecosystems,
biomes.

The book follows this sequence. Notice how it is the reverse of the classification hierarchy: the most inclusive levels here are at the bottom.

You will need to understand how processes operating at the level of individuals or populations can structure the community or ecosystem. These connections are made in Chapters 6, 7, and 8. This is important to demonstrate your capacity to see each topic in its widest context: answers that do this will score highly. This is why we encourage you to **Make the connection** throughout the book (Box 1.3).

Box 1.3 Make the connection

Integrating ecology and evolution

Flags linking related concepts appear throughout the text. This will help you to draw on information from different topic areas in your answers.

It is important that you are proactive in your revision, and look for these connections yourself. As a prompt, try the following exercise. Suggest how **ecological niche** might feature in an explanation of each of the following. Most require just one- or two-sentence answers.

Natural selection: reproductive success of a species

- Why should any species seek to occupy its fundamental niche?
- Why might a population confined to a marginal habitat eventually flourish?

Autecology: where and when a species occurs in a habitat

- Why do bluebells blossom in the early spring on the temperate woodland floor?
- Explain the difference in coat colour and pattern between lions and tigers.

Population ecology: realized niche and population growth

- Why should insectivorous birds defend territories in a temperate woodland?
- Why should population growth slow as the number of individuals reach the capacity of its habitat?

Community ecology: interactions between species

- Why do many herbivores create caches of food in a temperate woodland?
- Why should the ability of some species to colonize a habitat depend on which species have already arrived?

Ecosystem ecology: ecosystem functions

- Why should the loss of nitrogen from the soil increase when trees are felled in a temperate woodland?
- How might the age of an ecosystem determine its species diversity?

Hopefully, the answers to these should be obvious by the end of the book.

1.4 NOTICE

This chapter provides an overview of the entire revision guide. It should be useful for reviewing the main ideas in a short period of time.

Its most important function, however, is to help you establish the connections between the different sections.

Check your understanding

Examination-type questions

1. Suggest the important selective pressures in the following ecological niches.
   a. A stalking predator of large mammals
   b. An anteater
   c. Primary producer in deeper coastal waters
   d. Primary producer in a tropical grassland
   e. Parasite of caterpillars
2.
   a. What are the benefits to a predator of defending a large territory? What are the costs?
   b. Select a predator, specify its prey and say how you would measure these costs and benefits in the field.

You'll find answers to these questions—plus additional exercises and multiple-choice questions—in the Online Resource Centre accompanying this revision guide. Go to http://www.oxfordtextbooks.co.uk/orc/thrive/or scan this image:

2 Evolution by natural selection

Key concepts

- Organisms adapt to a changing environment—in the short term through phenotypic plasticity—over the long term by natural selection and genotypic adaptation.
- Homeostatic mechanisms maintain an organism's internal environment to preserve enzyme function and protein integrity.
- Each species has tolerance limits within which it can grow and reproduce. Outside this range the cost of living rises: energy and other resources are allocated between maintenance, growth, and reproduction. Here growth and reproduction may be checked by the high costs of maintenance.
- Some species survive a range of conditions. This phenotypic plasticity is typical of eurytopic species. In contrast, stenotopic species are closely adapted to particular environments and have a narrow niche breadth on key gradients.
- Niche overlap between species may indicate their competition. Character displacement may follow from both intra- and interspecific competition, as genotypes are selected to exploit a part of a resource gradient.
- The evolution of species follows from (i) the over-production of offspring and a competition for resources, (ii) the variation between individuals and

their differential success, whereby (iii) those most able to grow and reproduce succeed in passing on their genes to the next generation.

- Evolution results from persistent selective pressures over many generations, directional selection favouring some traits. Evolution has no purpose or endpoint and does not work to a design.
- Sexual selection can lead to exaggerated and apparently maladaptive traits, but which may secure a partner.
- The gene, the code for a trait, is the unit of selection. Its proliferation may be favoured by kin selection; that is, altruism favouring relatives carrying the same code.

Assumed knowledge

A basic understanding of homeostasis. The mechanisms of chromosome replication, mitosis, and meiosis.

The descriptions in this chapter apply principally to sexually reproducing organisms.

A conceptual map for this chapter is available on the companion website: go to http://www.oxfordtextbooks.co.uk/orc/thrive/ or scan this image:

2.1 ADAPTATION AND ACCLIMATION

Adaptation is a change in the biology of an individual that better fits it to its environment, improving its chances of surviving and reproducing. There are two types of adaptation: those that are acquired and those that are inherited.

- Acquired adaptations include the physiological, developmental, or behavioural changes made by an organism adjusting to conditions in its habitat.
- A plant wilts when it is short of water and a dog moves into shade to avoid the heat of the sun: both are homeostatic mechanisms to maintain the organism's internal environment. Such changes are reversible.
- Developmental adaptations may alter the pattern of growth—perhaps when food is in short supply or when leaves grow in deep shade—changes which may not be reversible.
- These are all phenotypic responses, rapid adjustments to new external conditions. The change itself is not written into its genes and cannot be inherited. This is called **phenotypic plasticity** or **phenotypic adaptation.**
- However, each response does have a genetic component: the individual will have inherited the *capacity* to make such adjustments.
- All organisms have some capacity for phenotypic adaptation because few habitats remain constant.

- Adjustments incur costs and a persistent change will favour genotypes that accommodate the new habitat with the least cost to their growth or reproduction.

Not all traits are adaptive: some characteristics are contingent. For example, a small body size may result from a poor food supply at a crucial stage in the development of the organism.

Genotypic adaptations are inherited. If a trait is favoured, or selected against, the genes coding for it change their frequency in the population. Genotypes which enhance fitness, and which are responsible for the most offspring, will dominate the next generation.

- Genotypic adaptation follows if the capacity for homeostatic or behavioural regulation is exceeded, or if these incur high costs.
- A persistent selective pressure causes the genotype of the population to change over generations.
- Variable environments select genotypes that can accommodate the most frequent or persistent pressures, favouring individuals with a high phenotypic plasticity.
- A rapid adaptability may not be advantageous in more constant environments, especially if this diverts resources from reproduction.
- Consequently, consistent ecosystems are populated by species closely adapted to their habitat, and which are less able to accommodate variable conditions.

Traits and genotypes that are selectively neutral will only change their frequency through chance.

➔ *Section 3.2, Evolution in the absence of selection*

Looking for extra marks?

Note that morphological species unchanged over geological time occur in habitats that are themselves unchanged: a long phylogenetic history can be indicative of an unchanged niche.

These may also have a high species diversity (consider the variety of adaptations to predation seen in deep-sea fishes).

➔ *Section 1.1, Time and the definition of a species; Section 8.2, Gradients of diversity*

Acclimation

- Protein structures are readily denatured and their function destroyed by extremes of temperature, acidity, and salinity. Conditions inside cells and tissues need to be maintained to protect enzyme function, structural proteins, and cellular processes.
- Acclimation is a short-term adaptation when external conditions are not optimal. It is a phenotypic change to maintain the internal environment.
- Organisms need these rapid and reversible responses to survive until conditions become more favourable, or they can move.

- Homeostatic mechanisms work harder when there are large differences between the internal and external environments, using more resources and incurring higher costs.
- Some organisms are highly adaptable and make major shifts in their physiology. For example, facultative anaerobes are microorganisms which thrive in oxygen-rich habitats but switch their metabolism when oxygen levels are low. In contrast, obligate anaerobic bacteria can only survive in the absence of oxygen.
- Mammals burn more energy and lose more water in environments hotter than their optimum. They can move to a cooler location (a behavioural response) but also have a range of physiological responses to accommodate high temperatures. Any response incurs costs, measured as the energy (or other resources) expended in staying cool.
- Some costs cannot be avoided: the resources needed to maintain cellular processes. The **basal metabolic rate** is the minimum energy expenditure needed to stay alive. Beyond this, and the energy required for resource acquisition, resources can be directed to **production** (growth and reproduction) (Figure 2.1).
- Natural selection weighs the benefits of growth against its costs, favouring an allocation of resources which maximizes the number of viable offspring.

  ➔ *Section 1.2, Cost-benefit analysis*
- Some species grow large and delay reproduction, while others grow little and reproduce early. A single, early reproductive effort typically results in many gametes but each at a small cost to their parents.
- The different strategies are determined by the predictability of a species habitat and the availability of resources.
- If a habitat (or its resources) is short-lived, only a short life cycle will produce offspring.
- Longer generation times are needed when more resources are needed, are limited, or take longer to acquire.
- Many habitats have a periodicity or seasonality to which organisms have to adapt. Their phenotypic plasticity accommodates these cycles, and schedules their resource acquisition, growth, and reproduction.
- For example, plants and animals living in tidal estuaries accommodate diurnal swings in salinity and exposure, and hydraulic forces of varying intensity and direction. Adaptability is a key trait for life in the estuary (Figure 2.2).
- Their morphology and other adaptations to this niche are the means by which we classify them into typological species.
- However, the small mussels found high up the estuary belong to the same species as the larger mussels close to the sea. They share much the same genotype but differ in their phenotypes.
- We might guess that the smaller mussels acquire less food or spend more energy on maintenance (or both) because of the longer period they are exposed at low tide.
- By transplanting juveniles between the two locations we might measure the relative significance of phenotypic and genotypic adaptation for their growth rate.

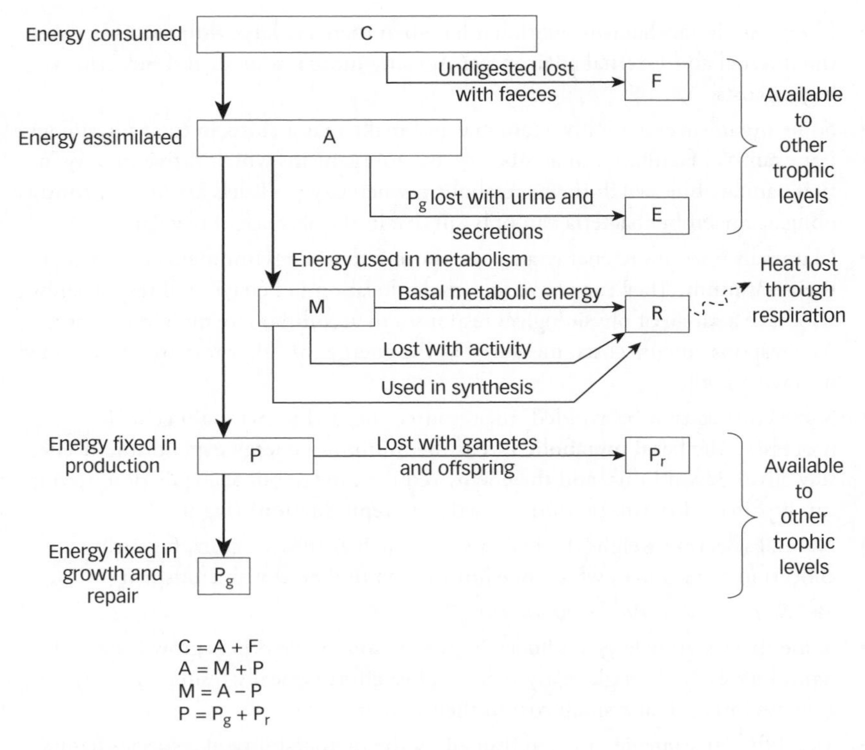

Figure 2.1 The partitioning of energy by an animal.

Only a proportion of the energy consumed (*C*) will be assimilated (*A*) and available to support metabolism (*M*) and production (*P*). Energy lost with the faeces (*F*) or in secretions (*E*) will be available to decomposers, whereas energy fixed in new tissues (*Pg* and *Pr*) can be passed on to higher trophic levels. (Note that *E* is often counted in *A* since this energy can be lost over the lifetime of the animal; also these boxes are not to scale.)

A large proportion of the energy is used to maintain basic functions, in locomotion, or in the cellular processes of anabolism and catabolism. This will be eventually lost as heat (*R*). All of these losses ($F + E + R$) mean that only a small proportion of the energy consumed will actually be available for predators feeding on this animal.

These equations work in the same way for an entire trophic level, and could be easily modified for the energetics of an autotroph.

Box 2.1 Make the connection

The energy budget

The energy budget presented in Figure 2.1 for an individual is equivalent to that of whole trophic levels (see Box 7.2 and Box 7.3). The energy lost here by the individual mirrors the inefficiencies described there as energy moves between trophic levels.

Notice how selective pressures on an organism (in this case to improve its energetic efficiency) could induce change in the structure and functioning of the larger community: rates of energy gain and loss by individual species shape its food web and trophic structure. This is one way in which the activity of a species can shape the niche of its neighbours.

Consider the significance of this for the integration of an ecological community: how might the loss of one species affect the larger community? Are some species more important than others? Is energy the only factor by which one species could impact the community?

➔ *Box 7.2, Box 7.3*

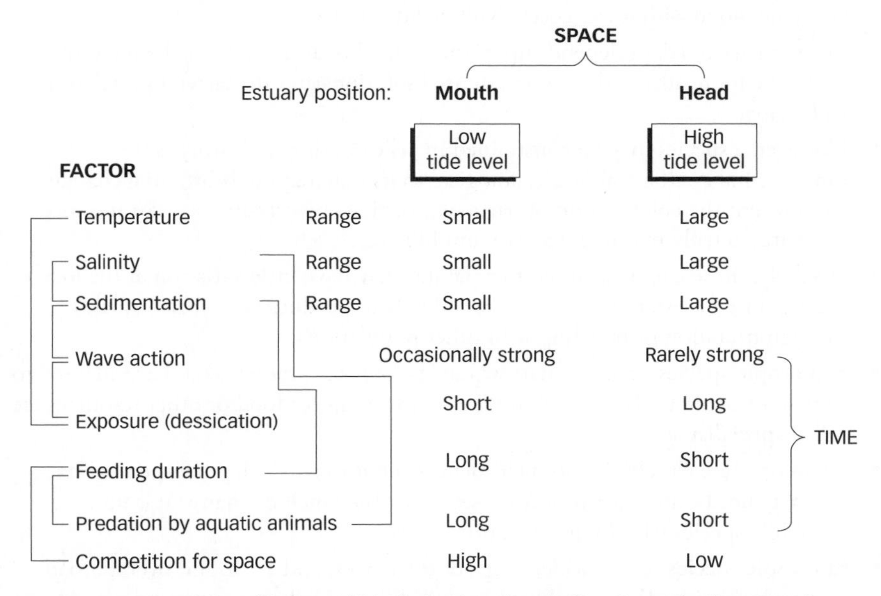

Figure 2.2 Some of the abiotic and biotic factors that define the niche for mussels (sedentary filter-feeding molluscs) inhabiting a tidal estuary. Both time and space determine the effect of each factor and produce the complex interactions (indicated by the connecting lines) which makes a full description of the niche difficult.
For the mussel, location is everything. It determines the duration of many of these effects: mussels situated far from the low-tide level experience longer periods of exposure (desiccation), shorter periods of feeding, and greater ranges of salinity and temperature. Conditions will also be less accessible to many aquatic predators. The hydraulic forces (wave action) a mussel must withstand also depend on time and position. The seasonality of the tides and freshwater flows through the estuary interact to affect food availability, growth, and osmotic stress. It also synchronizes the reproductive activity of the population.

Looking for extra marks?

Many organisms sequence their activity according to predictable cycles in their habitat: their migrations, feeding, and reproduction. In some cases, their niches may only be separated by time.

Have detailed examples for habitats you have studied.

Limits of tolerance and species types

- Demanding and persistent conditions in the environment will reduce the reproductive capacity of poorly adapted individuals.
- For example, conditions that impair the function of the common form of an enzyme may favour a structural variant, or **allozyme**, able to maintain its function.
- The selective advantage conferred by this allozyme means the allele coding for this form becomes more prevalent in the gene pool. In the same way a more efficient allozyme—one with lower costs—will be favoured.
- The capacity to adapt depends upon the variability at the loci for the relevant traits: the more alleles, the more potential solutions and the faster a population will change.
- However, a species may be constrained by its evolutionary history: some fundamental anatomical or physiological traits may require substantial change and prevent the colonization of particular habitats by certain taxa. For instance, there are no truly marine insects or amphibians.
- Similarly, those closely adapted to a habitat may have little variation at the loci coding for a necessary trait. They are unlikely to produce new alleles quickly through mutation or breeding with other populations.
- **Stenotopic species** occupy a narrow range along key gradients, and are adapted to consistent habitats. Those exploiting a narrow range of food or other resources are called **specialist species**.
- Stenotopes are closely tied to their niche, do not move readily between habitats, and may not be able to reproduce elsewhere. They include many large and immobile species with long generation times.
- **Eurytopic species** have a wider range of tolerances, and are found over a broad geographical area. They are able to exploit different habitats and resources. As **generalist species** they move freely to exploit what resources are available.
- Generalists on one environmental gradient have wide tolerance limits on most gradients. They are typical of changeable or unpredictable habitats. Most have short generation times with a dispersive stage in their life cycle.
- Few plants and animals are readily classified as perfectly stenotopic or eurytopic across all traits. However, the life history of many species is best explained by the consistency of the habitats or resources they exploit (see Table 2.1).

➔ *Figure 1.2, Limits of tolerance; Section 4.2, Reproductive strategies amongst* r- *and* K-*selected species*

Revision tip

Memorize the key characteristics of stenotopic and eurytopic species in Table 2.1. Try to relate each of these to the plant and animal examples you have studied.

	Character	r-selected	K-selected
Habitat features	Durational stability	Low	High
	Successional stage	Early	Late
Biological features	Body size	Small	Large
	Generation time	Short	Long
	Reproductive events	Few	Many
	Fecundity	High	Low
	Investment in each offspring	Small	Large
Population features	Sensitivity of birth rate to population density	Low	High
	Juvenile mortality rate	High	Low
Ecological features	Competitive ability	Low	High
	Efficiency of resource use	Low	High
	Dispersal ability	High	Low
	Investment in defence mechanisms	Small	Large

Table 2.1 The characteristics of eurytopic and stenotopic species, broadly equivalent to *r*-selected and *K*-selected species (see Section 4.2), respectively. *r*-selected and *K*-selected species are less precisely equivalent to generalist and specialist species. These labels are *not* used interchangeably by ecologists, but there is some equivalence between the categories. In each case population growth, resource utilization, and reproduction strategies can all be linked to habitat stability.

➔ *Section 4.2, Reproductive strategies amongst* r- *and* K-*selected species*

Box 2.2 Make the connection

Ecotypes and stenotopes

Some large and immobile species have populations closely adapted to local conditions, such that hybrids between populations are poorly adapted to either habitat. Examples include the different races of the African white rhino and some slow-growing temperate trees. Like many stenotopes, these are large-bodied organisms with long generation times.

Large individuals tend to form small populations because of the demands each individual places on space and other resources. With intense competition for these resources, specialist (and efficient) genotypes dominate. Because of their relative immobility their effective gene pool is small, with little exchange between populations. The result is highly localized races with distinct genotypes: ecotypes.

Note the connection here between body size, population size, life history, and reproductive strategy with the constancy of the habitat (Table 2.1).

➔ *Section 1.2, Ecotypes; Section 4.2, Body size and life-history strategies*

2.2 NICHE AND COMPETITION

- A full description of a niche would include all the gradients to which a species is adapted, including its interactions with other species (Figure 2.2).

Niche and competition

- Species may develop close partnerships with each other, from which both benefit. A partner will help define a species' niche and in some cases may actually be its habitat. Cooperation will persist as long as the costs to each partner do not exceed the benefits.
- Interacting species will co-evolve, adapting and adjusting to the conditions they create in each other's ecological space. Competitors adjust as they fight for a resource; predator and prey adjust to each other's behaviour.
- Niches flex together, and the structure of a community reflects the adaptations of its constituent species.

➔ *Section 5.5, Co-evolution; Sections 6.3, 7.3, Community structure*

Competition

Competition between members of the same species is **intraspecific competition.**

- Scarce resources may place an upper limit on the number of individuals a habitat can support.
- Selection then favours individuals most efficient in their use of a resource, able to maximize their reproductive success.
- Individuals can avoid competition and secure more of a resource by expanding into unexploited parts of a gradient (Figure 2.3).
- Natural selection thus drives all species to expand their ecological range.

Competition between species is **interspecific competition.**

- Where one species range extends into that of another there is **niche overlap**. This may indicate competition between two or more species.
- If one species is able to out-compete others, it will extend its range, occupying more of a resource. Others may then be confined to a smaller part of the gradient or lost altogether.
- The alternative is that each species becomes better adapted to a different part of the gradient. Each becomes more efficient in using its part of the resource, making it more difficult for other species to invade.
- Overlap may not check the growth of the species if the resource is abundant. Then there is no competitive battle, at least on this gradient.
- Alternatively, one species may not out-compete others on this gradient because its numbers are checked by a second (or third) factor. In this case, the competitive battle is never engaged.
- It is then possible, on some gradients, for the range of one species to lie completely within the range of another.
- Perhaps these battles were fought in the past so there is now no overlap on any gradient. The two species have already separated and become adapted for different parts of a resource spectrum (Figure 2.3).
- Where it does represent competition, a larger overlap indicates a greater selective pressure on both species.

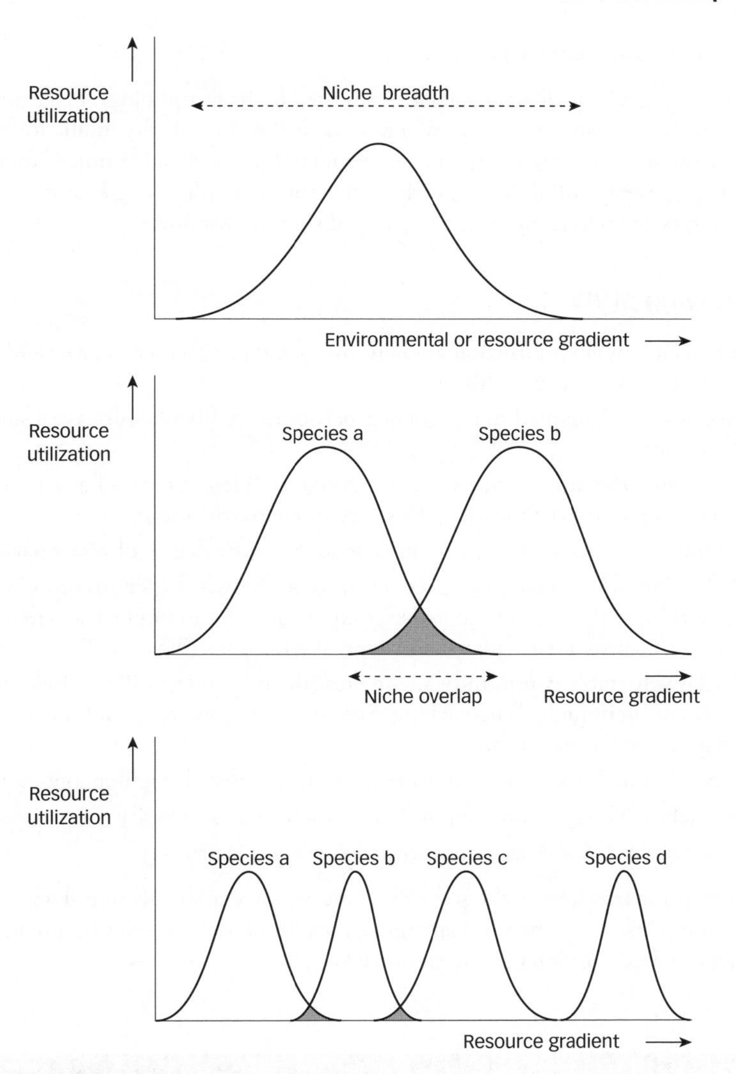

Figure 2.3 Measuring niche.

Niche measurements taken along a single environmental or resource gradient. 'Utilization' can be indicated by a variety of direct or indirect measures: the number of occurrences of a species, the amount consumed, the number of offspring produced, and so on.

(a) The range of an environmental gradient (such as soil pH) or a resource spectrum (such as prey size) that a species can exploit is its niche breadth.

(b) Competition may occur where two species occupy the same range; say, two plants competing to occupy soils with particular levels of nitrogen. The greater their niche overlap the larger the range over which they are competing and the more intense the competition between them.

(c) Species may be separated as a result of past competitive battles; through character displacement they become adapted to different parts of the gradient: this is niche differentiation. Adding more species incurs further displacement. This niche packing may be indicated by a smaller average niche breadth or a greater overlap.

Niche and competition

 *Section 5.2, Interspecific competition*

Often the key gradients for a species are indicated by its morphology, behaviour, or life history. For example, the shape of a mussel shell reflects the hydraulic forces it has to withstand; low-growing temperate woodland plants avoid competition for sunlight by growing and flowering early in the year, with physiologies and morphologies adapted to the shade of a closed summer canopy.

Quantifying niche

- The range of an environmental gradient over which a species can grow and reproduce is its **niche breadth**.
- A specialist, well adapted to a particular part of the gradient, will have a small niche breadth.
- Species using the same resource, each adapted to different parts of a resource spectrum, may co-exist together. This is **resource partitioning**.
- The number of species along a resource gradient is the degree of **niche packing.**
- This division of the ecological space means there is **niche differentiation** between species. Species that would otherwise compete may, for example, use a resource at different times, or a particular size or type of a resource (Figure 2.3).
- In a relatively stable community, niche breadth and overlap will decrease with time, resulting in high niche packing. Species have co-evolved, and niche differentiation has increased.
- We also find high niche packing in stable ecosystems with high-energy inputs.
- High niche differentiation is equivalent to high species diversity.

➔ *Sections 6.2, 8.2, Community structure and species diversity*

We can compare niche packing for the same resource gradients in different communities, just as we can compare niche breadth between species occupying equivalent niches in different ecosystems (Box 2.3).

Box 2.3 Key technique

Measuring niche

The measures used to quantify niche depend on the species and their key gradients. Darwin's finches are readily separated by the food they exploit and their beak morphologies; that is, their niche and their adaptations to this ecological space. Similarly, insectivorous birds in temperate woodlands can be divided by the size of their prey and where they feed in the canopy. Recording their consumption in each prey size category allows us measure both niche breadth and overlap between species on this gradient.

A variety of indices have been suggested, but at their simplest they include no measure of resource abundance. Two such measures are shown here.

Niche breadth can be measured by:

$$\text{Levin's index } B = \frac{1}{\Sigma p_i^2}$$

Where p_i is the proportion of individuals exploiting category i.

So, if most records for a species are confined to just two categories (say 0.4 and 0.6; $\Sigma p_i^2 = 0.52$) its niche breadth will be small ($B = 1.92$). If it occupies more categories (four; say 0.2, 0.3, 0.3, and 0.2; $\Sigma p_i^2 = 0.26$), B will be larger (3.85).

If the records were evenly distributed, B will equal the number of categories. Thus values of Levin's index below the number of categories measure the extent to which some categories are preferred over others.

Niche overlap can be measured by:

$$\text{Pianka's index } O_{ij} = \frac{\Sigma p_{ij} p_{ik}}{\sqrt{(\Sigma p_{ij}^2 \ \Sigma p_{ik}^2)}}$$

Where p_{ij} is the proportion of resource i used by species j and p_{ik} is the proportion of resource i used by species k.

Across five resource categories species j is found in the proportions 0, 0, 0.2, 0.4, and 0.4, so $\Sigma p_{ij}^2 = 0.36$. For species k these are 0.2, 0.3, 0.4, 0.1, and 0, so $\Sigma p_{ik}^2 = 0.3$.

$$\Sigma p_{ij} p_{ik} = 0.12 \text{ and } \sqrt{(\Sigma p_{ij}^2 \ \Sigma p_{ik}^2)} = \sqrt{(0.36 \times 0.3)} = 0.329$$

$$O_{ij} = 0.12/0.329 = 0.365$$

If species k is unchanged but species j occupies a larger range (0.1, 0.2, 0.2, 0.4, and 0.1), $\Sigma p_{ij}^2 = 0.26$, so now:

$$\Sigma p_{ij} p_{ik} = 0.20 \text{ and } \sqrt{(\Sigma p_{ij}^2 \ \Sigma p_{ik}^2)} = \sqrt{(0.26 \times 0.3)} = 0.279$$

$$O_{ij} = 0.20/0.279 = 0.717$$

Overlap between the two species has increased.

Niche overlap is equal to 1 when the two species have identical records.

Demonstrations of these indices are available in Worksheet 1 on the companion website; go to http://www.oxfordtextbooks.co.uk/orc/thrive/ or scan this image:

Revision tip

You need to illustrate niche packing, width, and overlap with examples in a community you have studied. A good example would show whether the degree of overlap is related to intensity of selection.

Intraspecific competition

- Individuals outside their optimum range will be under pressure to use resources more efficiently, to spend less on maintenance and more on production.
- If they can adapt to the marginal habitat they reduce their costs by avoiding competitive battles.
- Thus competition can lead to **character displacement**: genotypes adapted to different regions of a resource gradient.
- As groups separate they exchange genes less frequently with their neighbours and their gene pools become distinct.
- By adapting to soils with high levels of toxic metals, the **ecotypes** of some grasses avoid competition with other genotypes. They lose such battles when growing in uncontaminated soil: the costs of metal tolerance place them at a disadvantage in these habitats.
- Partitioning of a resource also happens between life stages of a species. To maximize its reproductive success an organism has to avoid competition between adults and juveniles. For example, dragonfly nymphs feed on insects at the bottom of the pond, whereas the adults feed in the air above it.
- We see the same differentiation in plants: species often have a **germination niche** distinct from that occupied by the adult. Many long-lived species produce seeds that remain dormant until their germination conditions are met.

Looking for extra marks?

Be aware of all the qualifiers that inevitably apply to real-world examples of niche overlap and packing.

How valid (discerning) are the measures used to quantify overlap or the intensity of competition? Do the data demonstrate competition for that resource? Might other factors affect the responses of either species? Is this competition current or past?

Niche and speciation

Character displacement subdivides a species into genetically distinct groups. Hybrids between these groups often have reduced fitness in either parental habitat: hybrid offspring of tolerant and non-tolerant grasses do not compete well with the parental populations growing in either the contaminated or the uncontaminated soils.

Speciation follows when the flow of genes between such groups ceases.

- With few reproductive exchanges, and with persistent selective pressures, each subpopulation begins to follow a different evolutionary trajectory.
- A change in these habitats may allow their gene flow to resume. Then the genetic differences will be diluted and perhaps lost.
- With little genetic exchange the two ecotypes become incipient species. With time they may become reproductively isolated from each other, unable to produce fertile offspring when mated together.

- The alternative is that one species out-competes the other and the loser goes extinct.

➔ *Table 3.1, Speciation and reproductive barriers*

2.3 EVOLUTION BY NATURAL SELECTION

Evolution is the process by which species change and originate.

- Species change over generations by the selection of inherited traits.
- New species arise when the accumulated changes prevent successful mating between populations.

Microevolution describes the changes which occur in the genetics of a population. **Macroevolution** describes speciation and the phylogeny of species.

Micro- and macroevolution have different time scales: microevolution operates over a few generations and may therefore be the subject of experimentation. Macroevolution describes the evolution of species through geological time. It is not amenable to experimentation in most multicellular organisms.

The principle

In Darwin's own words:

> Natural selection acts only by the preservation and accumulation of small inherited modifications, each profitable to the preserved being... [so that] ... new places in the natural economy of the district will be left open to be filled up by the modifications of the old inhabitants.
>
> *The Origin of Species*, 6th edn, Charles Darwin, 1872, Chapter IV

- This describes the occupation of a new ecological space, a new niche, and the character displacement that comes with adaptation.
- Darwin stated the principle of evolution by natural selection in very simple terms, as follows.
  - All species produce more offspring than their habitat can support and so there is a '*struggle for existence*'.
  - Individuals with an advantage, '*however slight*', would have a better chance of surviving than others.
  - '*Advantages*' inherited by the offspring might allow a species to change over generations and adapt to a new habitat.
- Although Darwin could not describe the means by which these 'advantages' were inherited, he knew from the breeding of domesticated plants and animals that artificial selection could drastically change an organism's appearance.
- He suggested that the accumulation of these inherited changes over many generations, without '*intercrossing*', allowed a population to diverge and become a distinct species.

Evolution by natural selection

By mutation

Change in chromosome structure

i. Deletion
ii. Inversion
iii. Insertion
iv. Translocation

Change in chromosome number

By rearrangement

At meiosis

i. Recombination
  a. At meiosis division 1
ii. Random assortment of
  a. Chromosomes at anaphase I
  b. Chromatids at anaphase II

At fertilization

iii. Gametes with different alleles fusing

Table 2.2 Sources of genetic variation.

- Speciation has occurred when the flow of hereditary information between populations is no longer possible (Table 2.2).

The elements of the principle

It is important that you can organize the elements of Darwin's argument into a coherent statement of the principle of evolution by natural selection, supported by examples.

- All species produce more offspring than can survive, so there is competition for limited resources.
- Those able to acquire these resources will survive and reproduce. The most efficient users of a resource will leave more offspring.
- Inherited traits that promote reproductive success will become more common.
- Over generations, this process of selection will fit a species closer and closer to the conditions in its environment.
- A divergent population, no longer able to breed with its neighbours, has become a new species.

Consider what the principle implies for our modern understanding of biology.

- With no variation between individuals there can be no selection or differential reproductive success other than by chance.
- If a habitat is unchanged, the same characters will be favoured from one generation to the next and there will be directional change in a trait. Less-favoured traits decline because of their relative reproductive failure.
- Phenotypes are selected, but it is information that is inherited. Only the information coded in the genes is passed on.
- Only genotypic variation expressed in the phenotype can be selected.

- In the natural world, chance and the environment determine which individuals reproduce. Chance factors will not consistently select one character and will not show **directional selection**: chance does not fit a species to its environment.

  ➔ *Section 3.2, Hardy–Weinberg equilibrium, Evolution in the absence of selection; Box 3.4, Measuring heritability*

Natural selection works with the variation available within a population and a population with limited variation offers fewer options. Few, if any, species can be described as perfectly adapted to their niche. They are, instead, continually adjusting (or failing to adjust) to the changing demands of their habitat.

- Chance can mean that the best genotypes fail to reproduce: traits that do not fit a species well to its niche may then dominate a population.
- The phylogenetic history of a species may constrain its capacity to adapt. The human back is one example. The kakapo, the burrowing parrot of New Zealand, is another: neither is well adapted to the demands placed upon it.
- In contrast, some species have an advantage colonizing new habitats because of **exaptation**. They have a trait, evolved under different selective pressures, that aids its colonization of a new ecological space. For example, feathers originally evolved as a form of insulation, rather than as an aid to flying.
- Offspring represent the success of the parent's genotype in the parental environment. They carry some of this code into the next generation but, in turn, their success depends on its fit to the habitat in which they find themselves.

Any character will only dominate for as long as the environment favours it or there is no selection against it. Evolution occurs when habitats change.

Box 2.4 Make the connection

Evolution by natural selection as an ecological principle

> Owing to this struggle, … variations, however slight, and from whatever cause proceeding, if they be in any degree profitable to the individuals of a species, in their infinitely complex relations with other organic beings and to their physical conditions of life, will tend to the preservation of such individuals, and will be generally inherited by the offspring.
>
> *The Origin of Species*, 6th edn, Charles Darwin, 1872, Chapter III

Not only is this a useful summary of the principle, it is also a statement of the ecological basis of natural selection, biotic and abiotic. The science of ecology was not recognized at this time but, with the publication of the *Origin* in 1859, it had its first theory. Today this theory is the central dogma of all biology: life fits itself to the ecological space available.

Ensure that you understand how the principle applies in others areas of biology you are studying. Draw on examples from your study of biochemistry, microbiology, and so on to illustrate it.

Adaptive evolution follows from directional selection

A consistent and persistent selective pressure will push a trait in a consistent direction. Much of our experimental evidence for evolution comes from selective pressures we have applied, where we can measure rates of adaptive change.

➔ *Box 3.4, Detecting selection*

The speed of adaptation can be slowed by several factors:

- a lack of genetic variation in the population, increasing the time taken for a advantageous allele to arise or become established;
- the dilution of such alleles in the population by gene flow from surrounding populations not subject to the same selective pressures;
- interactions between selective pressures for other traits: fitness may be increased by changes in one trait that compromises adaptation by another trait.

The most intense pressures are likely to induce the most rapid directional selection. However, adaptations may involve compromises between selective pressures, especially when a habitat changes rapidly, or following colonization of a new habitat. The fitness of an organism is determined by a range of selective pressures operating simultaneously.

A persistent selective pressure gives evolution a direction but not a purpose.

- Populations vary from one generation to the next, but without persistent selective pressures these variations have no direction.
- However, *direction is not purpose*: we see directional change in a trait or genotype when an adaptation persists from one generation to the next. But natural selection does not follow a blueprint for each species: it is blind. No purpose or endpoint is implied or required for a species to evolve.
- Co-evolution fits species together to form communities. Natural selection, acting on individuals, fits an assemblage of species, the community, to its ecological space.
- Again, the complexity and functioning of communities or ecosystems is not designed or directed. Their integration follows from each organism trying to do the same thing: securing the resources to maximize their reproductive success.

This ecological space, its niche, defines the species. However, this collection of selective pressures is subject to change and to chance.

Evolution explains phylogeny

Darwin's descriptor for evolution—'descent with modification'—implies that every species has a line of ancestors.

- Darwin recognized that his theory could explain the hierarchy of taxa and the classifications with which he was familiar. It could also explain the fossil specimens not seen in the living biota.
- This hierarchy implied the following:

- ◦ higher taxa are more inclusive, but their members have less in common;
  ◦ the lower the taxon, the more its members have in common and the more recently they have speciated;
  ◦ each taxon implies a common ancestor for its group;
  ◦ the most ancient taxa are morphologically simpler than more recent groups, in both the plants and animals.
- These implications are consistently reflected in the fossil record.
- Although there have been some surprises, many phylogenies based on the morphological species have been vindicated by studies of their molecular biology.

These analyses represent a major test of natural selection and the fossil record provides a record of descent with modification for the higher taxa of plants and animals.

➔ *Section 3.2, Population genetics*

The fewer traits they share, the more ancient the division between taxa: birds and bats share a common (four-limbed, walking) ancestor. This is phylogenetically closer than the ancestor they both share with insects, whose wings evolved from extensions of the exoskeleton.

- Different taxa arrive may arrive at similar solutions from different starting positions: this is **convergent evolution**. All modern birds (Class Aves) use an extension of their second and third digits of their forelimb to support their wing, whereas bats (Class Mammalia, Order Chiroptera) use the last four digits. Both are modified forelimbs originally used for walking, but have evolved independently.
- With **parallel evolution** species with the same ancestral trait but separated geographically arrive at similar solutions to the same selective pressures. Within the placental mammals, several large rodents in South America have similar morphologies to small grazers and browsers in various niches in the Old World tropics.
- Convergent and parallel evolution demonstrate that similar selective pressures produce equivalent adaptations.
- These morphological similarities can confuse our attempts at reconstructing an accurate phylogeny. Genomic analysis allows us to compare species' genetic code from indicative loci and establish their most likely ancestries. It would seem, for example, that the distinctive morphology of placental anteaters has evolved twice.

Revision tip

Memorize the essential elements for genotypic adaptation

Evolution occurs when inherited change is accumulated over generations. You should ensure that you understand the conditions under which this happens.

1. *The trait selected must be inherited.*

Phenotypic adaptations are lost with each generation. Only traits that are coded in the genes survive from parent to offspring.

2. *An individual carries the code for traits that were of selective advantage in its parent's environment.*

The parents were successful because their adaptations allowed them to reproduce...if the environment has since changed, their offspring may not be well adapted to the new conditions.

3. *One type dominates only while environmental conditions favour it, or there is no selective pressure against it.*

A trait can be retained, even if it is not adaptive, as long as it does not incur costs.

4. *Individuals need to differ for there to be selection.*

With no variation between them, all individuals have the same chance of surviving and reproducing.

5. *The gene or genes responsible for the adaptation were not induced by the environment.*

Genetic change is blind to the selective pressures in the environment. Novel genetic combinations arise through mutation or the chance combinations generated by sexual reproduction. They are not produced in response to the environment.

6. *The effects of natural selection are seen as changes in gene frequencies over the short term and by speciation events over the long term.*

In microevolution studies we can measure the fitness of a genotype by changes in its frequency from one generation to the next. Macroevolution events occur over many generations and much longer timescales.

2.4 NEODARWINISM

Neodarwinism is often referred to as 'the modern synthesis'. It combines the principle of evolution by natural selection with the details of the mechanism of inheritance.

- The mathematics of 'particulate' inheritance described by Mendel and the later work of population geneticists has allowed us to measure rates of change in the genetic code under natural selection.
- Mendel developed simple rules of inheritance based on discrete units, or genes. Morgan established that this information was carried on the chromosome, with a chemistry subsequently described by Crick and Watson. Even before the triplet code was described Dobzhansky, Fisher, Wright, and Haldane had begun to show how fitness could be measured in populations.
- These techniques allowed us to test Darwin's ideas, to measure the speed of adaptation, the intensity of selection, selective advantage, and inheritance.

➔ *Section 3.2, Hardy–Weinberg equilibrium*

Variation

Sexual reproduction is advantageous because it generates variation by producing new combinations of genes in different individuals (Box 2.5). Part of the genetic variation in a population derives from each individual having a pair of alleles at the same locus, one on each chromosome. If these alleles differ they add to this variation.

Additionally, some traits are coded at several loci, again each with a pair of alleles. This is polygenic variation, of which there are two types, as follows.

i. Discontinuous variation: a phenotype either shows a trait or it doesn't. There are just two (or a small number of) distinct and discrete phenotypes. The skin of peas in Mendel's experiments is one example, demonstrating the particulate nature of inheritance.
ii. Continuous variation has indistinct phenotypes which grade into one another. This may indicate that the trait has a large number of different alleles, is coded by genes at several loci, or there is co-dominance or an absence of dominance. Some flowers show a range of colours according to their combination of alleles, whereas in others colour depends on their soil. The former is genotypic variation, the latter is phenotypic variation.

In some cases, two different alleles at the same locus confer a **heterozygote advantage**, the most famous example of which is sickle-cell anaemia in areas where malaria is endemic.

Box 2.5 Reminder

Know the genetic basis of variation

Ensure that you understand the details below. This is necessary for the population genetics which follows later. (Consider writing two or three sentences on the biology behind each statement, or drawing a series of diagrams to check your understanding.)

A **gene** is a small piece of DNA, occupying a particular position (or **locus**) on a chromosome. It codes for the construction of a polypeptide, the building blocks of proteins. In diploid organisms with paired chromosomes there are two genes coding for the same character, one on the chromosome from each parent. Within a population there may be many alternate forms, or **alleles**, of that gene, although an individual can only have two. If these are identical the individual is **homozygous**; if different, it is **heterozygous.**

One allele may dominate over the other, although at some loci there is no dominance or there is co-dominance. Recessive alleles in heterozygous genotypes are not subject to selection but will add to the variation of the next generation. For the most part, any gene becomes subject to selection only if it (or its interactions with other genes) is expressed in the phenotype.

The selective advantage of sexual reproduction is that it preserves genetic variation through heterozygosity and generates phenotypic variation through

continued

recombination, **independent assortment**, and the random combination of gametes at fertilization. The chromosome on which the gene occurs, its locus, and proximity to other genes all determine its linkage to neighbouring genes and its interactions with the rest of the genotype. Within a population new genetic variation is introduced by immigration of new individuals and by mutation, a chemical change in the genetic code. Not all mutations lead to a different polypeptide because there is a high degree of redundancy in the triplet code used in **translation** and **transcription**.

Before division, genetic material in a cell is replicated. In somatic tissues (cells other than the sex cells) there is one division of the replicated chromosomes, **mitosis**, which pulls apart the sister chromatids. This preserves their **diploid** state at division, and the daughter cells are genetically identical.

In the sex cells replicated chromosomes undergo two divisions, firstly separating homologous chromosomes and later the sister chromatids. This is **meiosis**, leading to **haploid** gametes. Because of recombination and independent assortment at meiosis, the gametes are genetically different from each other. This, together with their random combination at fertilization, is how sexual reproduction produces zygotes that are genetically unique.

Nature and sources of genetic variation

Species which range over several habitats or a wide range of environmental gradients can show a variety of phenotypes. Different genotypes are favoured in different habitats, and we can sometimes correlate variation in a trait with environmental gradients. A directional change in a trait over time may also be indicative of adaptive change.

Alternatively, it may have no evolutionary significance: it may be a phenotypic (acquired) adaptation that is not inherited. Populations can also show genetic variation that is not adaptive but persists by chance.

- Variation in the genes which does not affect the fitness of the individual is called **neutral variation**.
- Chance processes in genetics, like recombination, can produce heritable changes that may persist in a population *without* being favoured by selection. They are not lost because they are not selected against.
- Some traits which determine fitness in one habitat have no significance in a different environment.
- Only traits which correlate with reproductive success are likely to be adaptive. These have to persist over generations to demonstrate they are inherited and being selected.
- A positive correlation would suggest the trait is favoured; a negative correlation would suggest it is being selected against.

Within a population there are two main sources of variation (Table 2.2):

i. mutations producing entirely new code,

ii. rearrangement of existing code.

The latter follows from chromosome replication, from chromosome movement during meiosis, and new combinations of alleles formed at fertilization (Box 2.5).

Additionally, new code can be introduced by

iii. gene flow,

when mating occurs with individuals from neighbouring populations.

- Rapid adaptation occurs when the trait under selection has high **heritability** and also a high variability. That is, most of the variation is genotypic and results from a large number of alleles associated with each locus.
- When selective pressures are high there will be large differences in the reproductive success between individuals. Favoured traits will be more common in the next generation—they will have high heritability—and selection will produce rapid directional change.

➔ *Box 3.4, Measuring heritability*

2.5 TYPES OF SELECTION

Any environment will demand fundamental adaptations to its particular physics and chemistry: the adaptations needed to move through water or fly through the air, for example. Additionally, selective pressures result from a species' interactions with other species or with other individuals of the same species.

Selection can also be classified by its effect on the gene pool. We may observe no variation in a trait because it has become fixed in the population due to:

- **stabilizing selection**, where the greatest fitness is close to the current mean of a trait and extreme phenotypes have a lower relative fitness. This indicates that a population is well matched to existing conditions in its habitat.

We may observe change when there is:

- **disruptive selection**, when an extreme phenotype enjoys higher fitness because of its difference from the population mean. In this case, changed conditions in the habitat may favour a different phenotype. Extreme phenotypes are possible because recessive genes in heterozygotes are sometimes expressed in homozygous offspring. Disruptive selection can cause populations to show several **polymorphisms**: this may be advantageous if the different morphs do not compete with each other.

Or alternatively when there is:

- **frequency-dependent selection**, when rarity itself confers an advantage; for example, when a predator needs to learn a 'prey image', and being rare makes a prey less easily identified. This advantage disappears when the frequency of the rare phenotype increases.

Types of selection

In each of these cases we observe:

- **directional selection**, where individuals with genotypes different from the current population mean have a higher (or lower) relative fitness. Selection has a 'direction' because the frequency of the genotype changes consistently over generations.

Revision tip

Ensure that you have detailed examples of stabilizing, disruptive, and frequency-dependent selection, in both plants and animals, related to selective pressures in their ecological space. You will also need examples of sexual selection, and especially assortative mating.

Sexual selection

Darwin recognized that selecting a mate was a special form of selection, which might explain morphological differences between the sexes (sexual dimorphism) as well as the apparently maladaptive traits of some displays or behaviours.

- **Sexual selection** results from inherited differences in the ability to obtain a mate, arising from competition between members of the same sex. Darwin contrasts this with natural selection, writing that the outcome of such competition was 'not death…but few or no offspring'.
- Sexual selection does not just operate on males. The gender incurring the largest costs in reproduction usually gets to choose: male pipefish guard the brood and make the largest investment in the protection of the young. Males get to choose the females.
- Competition between potential partners can lead to investment in displays, aggression, and territoriality. The rewards differ: a large territory, defended from competitors, may allow a male to maintain a harem, yet a nest may only secure a single female.
- The peacock's tail is another flag of male fitness. Elaborate dances and bright colours are also signals of the quality of the owner's genotype and health, despite the risks or costs they also represent.
- Such signals of fitness indicate that a potential father carries the genes to attract partners, defend large territories, or build good nests, traits which the female would want her male offspring to have: some indication that her genes are likely to be passed on by her offspring.
- When one gender favours partners that match its own genotype, called **assortative mating**, these characters become reinforced. A female with a gene that favours a particular feature in males—such as long tail feathers or horns—will produce offspring carrying both the genes for the preferred male trait *and* the gene for this preference. This creates a positive feedback and the rapid evolution of extreme phenotypes.
- This can be a form of directional (and disruptive) selection by reinforcement, where the more extreme traits enjoy greater fitness: 'runaway' sexual selection.

- Sexual selection may also maintain differences between populations, preventing gene flow between different polymorphs. Without the correct signals, mating between these phenotypes may cease. This is one reason for the great diversity of cichlid fishes in the African Rift Valley lakes and the fruit flies of Hawai'i.

Altruism and kin selection

Sacrificial or altruistic behaviour was recognized by Darwin as problematic: how might a behaviour that reduced the fitness of an individual persist in a population subject to natural selection?

Such behaviours occur in the social insects (ants, bees, termites), other arthropods, and several classes of vertebrate. The modern synthesis provides an explanation for this **altruism**, based on shared genetic identities.

- We can perhaps understand why a bird, defending three or four eggs in a nest, might feign death to distract a predator from stealing its eggs.
- The cost to the parent is the risk of losing the opportunity to produce further eggs. Yet this trait is seen repeated over generations (and in different species) so it must confer a selective advantage.
- An advantage means the costs to the bird are outweighed by the benefits to its eggs. But then how do we compare these costs and benefits across several individuals, the parent and its offspring?
- This is possible if the gene is the unit of selection. Now the focus is not the individual but the genes shared by the parent and offspring. The cost-benefit analysis is now for those genes which govern this behaviour.
- This was the major insight of W. D. Hamilton and others: only genes are inherited and genes with traits that improve their own transmission will be favoured by selection.
- The bird incurs the risk because its eggs share many of its genes, including (probably) those prompting this behaviour. If such a gene increases its representation in the next generation, natural selection will have favoured such altruism.
- This principle is termed **kin selection** and extends to relatives other than offspring. The cost-benefit equation works exactly as before, with relatives making sacrifices for each other, for the benefit of their shared genes. Adaptations that increase the reproductive success of their genotype ensure that genes promoting kin progress to the next generation.
- We can measure this as the **inclusive fitness** of an individual. An individual can improve the chances of its genes passing to the next generation not only by its own reproductive efforts (its direct fitness) but also by aiding those of relatives likely to share its genes (its indirect fitness) (Box 2.6).
- Inclusive fitness is thus an organism's *individual fitness* (the number of its direct descendants) plus *the fitness of the recipients of its altruism, weighted by their genetic relatedness* (the weighted number of indirect descendents).

Types of selection

- Altruistic acts can be seen in different animal groups, especially when opportunities to reproduce are small: individuals with low direct fitness can increase their inclusive fitness by being altruistic (increasing their indirect fitness). For example, if all territories are occupied, the offspring of some birds forgo reproduction to help their parents raise their siblings and half-siblings (Box 2.6).

This is an important element of neodarwinism and begins to explain the evolution of social behaviour among animals. It also explains why, among the social insects, most individuals make no attempt to reproduce at all.

Box 2.6 Key technique

Calculating inclusive fitness

The chance of a gene passing to the next generation is the sum of the reproductive success of all individuals that carry it. This is the fitness of the gene, rather than that of the organisms: a gene promoting altruism will increase its own fitness if it raises the inclusive fitness of its carriers.

Fitness for the organism is measured as its **fecundity**, the number of offspring it is able to produce. Inclusive fitness is the sum of its own reproductive output (its direct fitness) and the reproductive output of relatives that share its genes (indirect fitness). Incurring costs for the benefit of near relatives, increasing their reproductive success, may therefore increase the inclusive fitness of the donor. The more closely a recipient is related to the donor, the more genes they are likely to share, and the larger the contribution an altruistic act will add to the donor's indirect fitness.

Hamilton created a rule to predict when altruism would be favoured by inclusive fitness:

$$r > c/b \text{ (Hamilton's rule)}$$

or

$$r \times b - c > 0$$

where:

- r = relatedness between two individuals (based on their genotypes),
- c = costs to the donor in reduced fitness (lower reproductive success),
- b = benefits to the recipient (increased reproductive success).

Simply, altruism will be favoured when the cost to the donor is low compared to the benefit to the recipient, according to their relatedness, r.

Consider an altruistic act for a full sibling, where donor and recipient share the same parents ($r = 0.5$). When the benefit to the recipient is an increase in fecundity of 30, and the cost to the donor is a reduction in fecundity of 10 then:

$$r \times b - c = 0.5 \times 30 - 10 = 5$$

So, the numbers of those with the gene promoting this altruism will increase as a result of the altruism. Alternatively, when the recipient is a half-sibling (sharing only one parent, $r = 0.25$):

$$r \times b - c = 0.25 \times 30 - 10 = -2.5$$

Here, the opposite is true.

This rule is very simple, and does not allow for complicated interactions between species. It also assumes a relatively stable population. However, there is now considerable experimental evidence to support its principle, even among humans.

Group selection

Living in a group can confer benefits (or reduce the costs) to its members, including those that do not share the same genotype. Groups of grazing or foraging mammals take turns watching for predators while most feed. In the extended families of ground squirrels or meerkats, this behaviour is explained by the inclusive fitness of individuals, but this would not apply in large herds of zebra or antelope, where the relatedness between some individuals would be small.

- Part of the benefit comes from the reduced costs of maintaining a collective vigilance, as long as each individual takes their turn on guard duty.
- In its most general form, **group selection** does not require individuals within a group to be related, only share the same behaviours. A gene that invokes cooperation, it is suggested, will raise the fitness of the group, and this group would flourish over other groups that had little cooperation.
- The benefit to the individual is indirect, deriving from the persistence of its group.
- This social behaviour is a property of the group, but to be inherited it has to be coded in its individual members. Because the advantage is indirect and needs a large group to work it is generally regarded as a weak selective pressure.
- Such behaviour is vulnerable to cheats. A gene for cheating would increase the fitness of its owner at the expense of the non-cheats.
- Cheats not sharing the task can continue to feed, incurring none of the costs of vigilance. As more individuals cheat, the risks of being predated rise and eventually the group may suffer. Cheating works only if most do not cheat.

Group selection remains highly contentious, not least because individual fitness (and the selection favouring these genes) is some distance from the differential selection of groups.

- The 'social trait' must confer an advantage on individuals in groups dominated by non-cheats, resulting from the superior performance of the group, to allow this genotype to become dominant.
- Any effect of group selection is believed to be minor because the variation in fitness between groups is likely to be less than that between individuals, especially

where there is gene flow between groups. Consequently low levels of gene flow between groups and low mutation rates would be necessary.

- Some have argued that group selection can be treated as an extension of inclusive fitness, and that it explains the extreme social behaviour (**eusociality**) of some insects.

Check your understanding

Examination-type questions

1. Describe how the main selective pressures on mussels inhabiting a tidal estuary depend on:
   a. their position,
   b. their interactions with each other.
2. Devise an experiment that could measure the relative contribution of phenotypic and genotypic factors in the average size of mussels taken from the top and bottom of their range within a tidal estuary.

You'll find answers to these questions—plus additional exercises and multiple-choice questions—in the Online Resource Centre accompanying this revision guide. Go to http://www.oxfordtextbooks.co.uk/orc/thrive or scan this image:

3 The evolving population

Key concepts

- Species are not always clearly demarcated in nature. Some plant groups have poorly defined boundaries and breed in aggregates of morphological species.
- We can detect and measure evolution as a persistent change in the allele frequencies of a population over generations. This may follow from consistent selective pressures, and has to be distinguished from chance effects.
- Genotypes may diverge when a small number of individuals establish a breeding population due to:
  - a founder effect: when a new population does not reflect the variation of its parent population;
  - genetic drift: when reduced gene flow allows isolated groups to develop distinct genetic identities;
  - genetic fixation: when an isolated population loses all variation at one or more loci.
- Under certain conditions, the Hardy–Weinberg equilibrium serves as a null hypothesis for selection. Over several generations, it can establish whether there is directional selection.
- Changes in allele frequencies allow us to measure the intensity of selection (coefficients of selection) and the response of the population.

continued

- We can measure the heritability of a trait and attribute phenotypes to genetic and environmental variation.
- The neutral theory of evolution suggests that genetic drift explains much of the non-adaptive change in DNA sequences and protein forms.
- Reproductive isolation defines a new species. A variety of pre-zygotic and post-zygotic barriers limit gene flow between populations.
- Allopatric speciation results from geographical isolation; neighbouring species adapt to different niches in sympatric speciation; parapatric speciation as a special form of sympatric speciation among large and immobile organisms.

Assumed knowledge

Basic genetics and the different forms of genetic and chromosomal mutations.
A basic understanding of sampling in statistics.
A conceptual map for this chapter is available on the companion website; go to http://www.oxfordtextbooks.co.uk/orc/thrive/ or scan this image:

3.1 FROM POPULATION TO SPECIES

Natural selection dictates which individuals reproduce but it is populations that change, and may evolve into new species. A population acquires a distinct genotype when mating occurs infrequently with neighbouring populations. When these exchanges no longer produce viable offspring, or do so only rarely, a new species has been created. Speciation is therefore the study of population genetics.

Types of species

Genetic distances between taxa are greater higher up the taxonomic hierarchy. However, this hierarchy is a ranking of taxa, and distances between them are not consistent across lineages.

Systematics has to reflect the important differences between taxa and the genetic distance between species may be small. Groupings below the species level make finer distinctions, as described here.

- **Demes** are inbreeding groups within a population, typically a local subpopulation. They have distinct genotypes, but there is no barrier to their breeding with surrounding populations.
- **Semi-species** are groups that have limited gene flow with other groups in the population, perhaps through geographical isolation. These are roughly equivalent to **ecotypes** or **evolutionarily significant units** because they are distinguished by their local adaptations, and have a different evolutionary trajectory from their

neighbours. They may eventually form new species but currently will produce fertile hybrids with their neighbours.

➔ *Section 1.1, Ecotypes and ESUs*

Many plant groups do not have distinct species.

- A **syngameon** is a collection of morphologically differentiated species found in some plant genera. They can interbreed with each other and produce fertile hybrids.
- Some have forms that appear as distinct species. These may be apomictic or **microspecies**: grown from seed produced without fertilization. This is a form of asexual (clonal) reproduction. However, most are able to reproduce sexually and will swap genes with each other.
- Often a species consists of several **apomicts**, which are then grouped together as **aggregates**.

Hybridization is far less common in the animal kingdom.

For most purposes, the 'species' can used as a functional definition, when gene flow between congeners is either impossible or highly restricted.

3.2 POPULATION GENETICS

We detect evolution, and measure its rate, by looking at changes in the gene pool of a population. But we have to distinguish the random changes associated with sexual reproduction, and other sampling effects, from the consistent changes over several generations that could indicate selection.

- Not all alleles in the gene pool contribute to the next generation. Only a fraction of the gametes create a zygote and only a fraction of these grow into fertile adults. For this reason, the gene pool is said to be sampled.
- As in statistics, a representative sample accurately reflects the population from which it was taken, with the same allelic frequencies. Small samples are less likely to be representative.
- Sampling happens when a few individuals colonize a new habitat. A **founder effect** is the creation of a distinct genetic group when a small number of colonists are sampled from the parent population.
- With little gene flow between the populations, future generations can maintain or exaggerate these differences.
- Small gene pools diverge rapidly from the parent population (Figure 3.1): rare alleles are lost easily because few individuals carry them and they have less chance of being sampled in each round of sexual reproduction.
- Conversely, chance might cause rare alleles to be over-represented in a sample, although this is less likely. Either way, the frequencies change rapidly in a small population, a process called **genetic drift**. Isolated, a small gene pool becomes genetically distinct over a few generations.

Population genetics

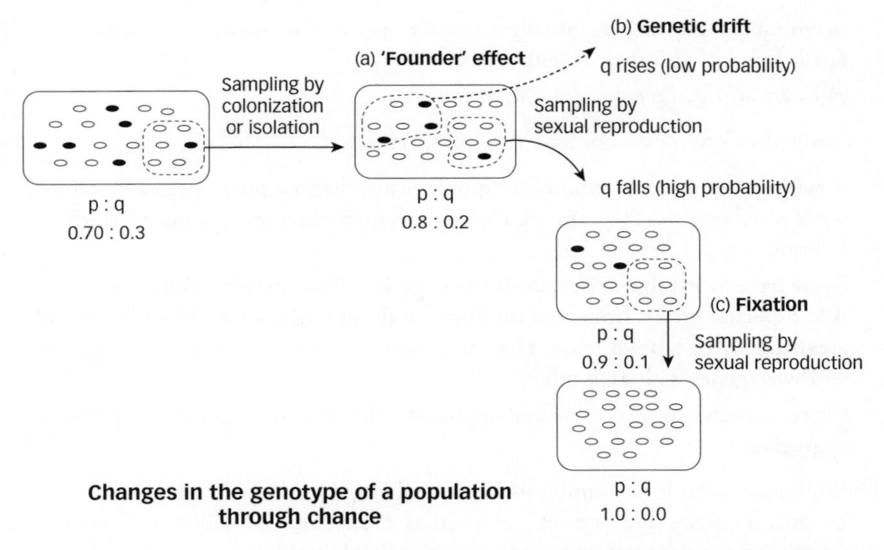

Figure 3.1 Chance can be an important factor determining the genotype of a population, especially when that population is small.

Allele frequencies are illustrated here by the filled and open symbols.

The founder effect (a), genetic drift (b), and fixation (c) all result from a small sample of the gene pool: (a) first in the process of colonizing a new habitat; (b) second in sexual reproduction, when only a few gametes produce zygotes, which, if repeated over generations, leads to fixation (c).

- With fewer exchanges with the parent population, divergent gene pools may eventually no longer breed with the parent population. This divergence may not be due to selection, but a result of sampling effects.
- Ultimately, the loss of rare alleles can lead to **genetic fixation**, when all individuals share the same allele at some loci. Fixation negates the advantages of sexual reproduction and only migration or mutation can then introduce variation at those loci.
- Smaller populations provide fewer opportunities for mutation. Even in large populations, mutation rates are too low to drive evolution, but they can be an important source of variation.
- Such sampling effects promote speciation and the barriers that can prevent gene flow below populations (Table 3.1).

The differences between populations can be diluted by significant gene flow between them. This is called **admixture** and can be an important source of genetic novelty, especially for populations that have been isolated for some time before this exchange resumes.

- Many organisms ensure gametes are swapped between populations—when individuals move between animal populations, or when pollen is produced by flowers before their own ovules are receptive, ensuring cross-pollination.

Pre-fertilization (pre-zygotic barriers)

Ecological isolation
Populations are separated by distance or barriers.

Temporal isolation
Populations are reproductively active at different times.

Behavioural isolation
Males and females of different populations fail to recognize each other as potential mates.

Mechanical isolation
Reproductive organs fail to match and the gametes never meet.

Gametic isolation
The sperm and the egg do not recognize each other so there is no fertilization.

Post-fertilization (post-zygotic barriers)

Hybrid inviability
The embryo fails to develop so the hybrid never reaches the adult stage.

Hybrid sterility
Offspring are infertile.

Hybrid breakdown
The offspring produce young that fail to develop or reproduce, or are poorly adapted to their habitat.

Table 3.1 Speciation and reproductive barriers.

- Genetic variation itself must confer some advantage for these strategies to persist. **Heterozygote advantage** follows from some increase in fitness from having more than one allele represented at several key loci. This has been described for certain traits in various groups.

   *Section 2.4, Neodarwinism; Table 2.2*

Natural selection *does not* favour variation (or any other trait) for the 'good of the species': genetic variation must confer an advantage on the individual if mechanisms that promote it are to persist in a population.

Revision tip

Ensure that you have examples of the strategies used by plants and animals to promote sexual reproduction and genetic exchange. You should also have a detailed example of heterozygote advantage, demonstrating the adaptive advantage of genetic variation at particular loci.

You also need examples for each type of pre-zygotic and post-zygotic reproductive barriers (Table 3.1).

Only genotypic variation expressed in the phenotype can be selected. This is sometimes referred to as the **transcriptome**: the expression of the genome represented by an individual's phenotype.

Looking for extra marks?

Note an important point here: natural selection works on individuals, and it is through differences in individual fitness that populations change. The previous chapter reviewed kin selection and altruism, where the effect of selection has an impact beyond the individual. Always be clear that selection can only work on inherited code, and the genes carried by individuals.

➔ *Section 2.5, Kin selection*

Hardy–Weinberg equilibrium

Revision tip

It is important that you understand the principle behind the Hardy–Weinberg law and can demonstrate this by completing an example calculation. Ensure that you follow each step in the examples given.

Note especially the cycle in each calculation: from the proportion of alleles in the haploid gametes to the proportion of genotypes in the diploid adults and then back again.

The **Hardy–Weinberg law** enables us to detect selection in a population as a persistent change in the proportion of alleles or genotypes. Under particular conditions (set out under Terms and conditions), **and in the absence of selection**, both proportions should remain **unchanged** from one generation to the next. Departure from these equilibria may indicate selection.

In the absence of a selective pressure

Consider a sexual diploid plant or animal. Every individual has two alleles for each gene (which may be the same), one on each chromosome. Although there may be many different alleles in the population, in our simple example there are just two possible alleles at this locus: A_1 and A_2.

➔ *Box 2.5, The genetic basis of variation*

Gamete allele frequencies

Thus, all the gametes produced by an adult are either A_1 or A_2. Let us say that 60% are A_1 and 40% are A_2:

$$\text{Proportion of } A_1 \text{ gametes } p = 0.6$$
$$\text{Proportion of } A_2 \text{ gametes } q = 0.4$$

Note that $p + q = 1.0$ (all gametes are either A_1 or A_2).

After fertilization, the possible combinations in the diploid zygote are:

$$A_1A_1,\ A_2A_2$$
$$A_1A_2,\ A_2A_1$$

Both genotypes on the first row are homozygous for this locus; those on the second row are heterozygous. The two heterozygote genotypes differ only by their source of either allele (sperm or egg).

Genotype frequencies in the diploid organism

Given the proportions in the gametes we calculate the frequency of each genotype when they combine to form the zygotes:

	Genotype	*Frequency*
Homozygous A_1	A_1A_1	0.6×0.6 (p^2) $= 0.36$
Homozygous A_2	A_2A_2	0.4×0.4 (q^2) $= 0.16$
Heterozygous	A_1A_2	0.6×0.4 (pq) $= 0.24$
Heterozygous	A_2A_1	0.4×0.6 (qp) $= 0.24$

(The frequencies of the two alleles are multiplied together because each gamete is selected randomly and *independently* of each other. So the chance of an A_1 sperm fertilizing an A_1 egg is the product of their two frequencies: each has a probability 0.6 of being selected at random, so $0.6 \times 0.6 = 0.36$.)

Collecting together the two heterozygotes, we have $(pq) + (qp) = 2(pq)$. The sum of all genotype frequencies in the *zygotes* is:

$$p^2 + 2(pq) + q^2$$

That is:

$$0.36 + 0.48 + 0.16 = 1.0$$

The entire population is partitioned between these three genotypes in these proportions. Thus 48% of the zygotes (and eventually adults) will be heterozygous at this locus.

If there had been selection—perhaps one genotype suffered greater mortality—we could not predict their proportions from the frequency of alleles in the population, and this equation would not work. This would be a departure from the previous balance of genotypes, and an indication that selection might have taken place.

At meiosis, these alleles separate again to form gametes. A homozygous adult will produce either two sets of A_1 or two sets of A_2 gametes; half of the gametes from a heterozygous adult will be A_1 and the other half A_2.

So, of all the gametes, 36% (from homozygous adults) and 24% (from heterozygous adults) will be A_1. You should be able to work out the proportion of A_2 gametes.

Gamete allele frequencies

The gametes in the gene pool from these adults will thus be given by:

$$\text{Proportion of } A_1 \text{ gametes} = p^2 + \tfrac{1}{2}(2pq)$$
$$= 0.6^2 + \tfrac{1}{2}(2 \times 0.6 \times 0.4) = 0.36 + 0.24 = 0.6$$

$$\text{Proportion of } A_2 \text{ gametes} = q^2 + \tfrac{1}{2}(2pq)$$
$$= 0.4^2 + \tfrac{1}{2}(2 \times 0.6 \times 0.4) = 0.16 + 0.24 = 0.4$$

There has been no change in the frequency of the two alleles.

The next generation will have the same proportions of the three genotypes. With complete segregation of the genes, and no selective pressure, the proportions of the two alleles, and the resultant genotypes, remain unchanged from one generation to the next.

In the absence of a selective pressure, this equilibrium will be maintained down the generations. This is called the Hardy–Weinberg equilibrium and represents a ***null hypothesis for selection***. If there is selection for a genotype (under the particular conditions set out below), the proportions of either the alleles or the genotypes will change.

In the presence of a selective pressure

We can see this by applying a selective pressure against a_2. We will also say that this allele is completely recessive (denoted now by its lower case letter) to A_1. Thus, only one genotype is selected against—a_2a_2—***when this allele is expressed in the phenotype***. Half of those with this phenotype die without reproducing and adding their gametes to the gene pool.

For simplicity we start with 100 individuals in the population:

Genotype		*Frequency*	*Numbers*
A_1A_1 =	0.6×0.6	= 0.36	= 36 individuals
A_1a_2 =	$2(0.6 \times 0.4)$	= 0.48	= 48 individuals
a_2a_2 =	0.4×0.4	= 0.16 minus (16 × ½)	= 8 individuals
Total = 92 individuals			

Notice that now we cannot predict the proportions of each genotype based on the starting values of p and q: there is a departure from the Hardy–Weinberg equilibrium because of differences in the death rates between the genotypes.

Each surviving individual represents two alleles:

$$\text{Homozygous } A_1 : 36 \times 2 = 72\ A_1$$
$$\text{Heterozygous} : 48 \times 2, \text{ which are } 48\ A_1 \text{ and } 48\ a_2$$
$$\text{Homozygous } a_2 : 8 \times 2 = 16\ a_2$$
$$\text{Total} = 184 \text{ alleles}$$

Note that 16 a_2 alleles were lost with the death of eight homozygous a_2 individuals.

Gamete allele frequencies

Of the 184 alleles:

$$72 + 48 \text{ are } A_1 : 120/184 = 65\% \text{ or } 0.65$$

and

$$16 + 48 \text{ are } a_2 : 64/184 = 35\% \text{ or } 0.35$$
$$p + q = 0.65 + 0.35 = 1.0$$

Genotype frequencies in the diploid organism

So the proportions of the three genotypes in the next generation will be

$$p^2 + 2(pq) + q^2$$

That is:

$$0.42 + 0.46 + 0.12 = 1.0$$

And (for completeness), in the gene pool of the next gametes:

Gamete allele frequencies

$$\begin{aligned}\text{Proportion of } A_1 \text{ gametes} &= p^2 + \tfrac{1}{2}(2pq)\\ &= 0.65^2 + \tfrac{1}{2}(2 \times 0.65 \times 0.35) = 0.42 + 0.23 = 0.65\\ \text{Proportion of } a_2 \text{ gametes} &= q^2 + \tfrac{1}{2}(2pq)\\ &= 0.35^2 + \tfrac{1}{2}(2 \times 0.65 \times 0.35) = 0.12 + 0.23 = 0.35\end{aligned}$$

As long as mating with these gametes is entirely random, a new Hardy–Weinberg equilibrium will be established with these new frequencies…and will persist unless there is further selection.

Finally, note that genotype proportions may change even if the allele frequencies remain the same; for example, when homozygotes are at a selective disadvantage compared to heterozygotes. Then, if both homozygotes suffer the same selective pressure, both alleles are reduced but their frequencies in the population remain unchanged. Here heterozygotes have a selective advantage, and the proportion of this genotype rises in the adult population.

Box 3.1 **Looking for extra marks?**

Since $p + q = 1$, then $q = 1 - p$

We could have got there a lot faster. To demonstrate the derivation, consider A_1 alone:

$$\text{The proportion of } A_1 \text{ in the gametes} = p^2 + \tfrac{1}{2}(2pq) = p^2 + pq$$

Substituting $1 - p$ for q:

$$\begin{aligned}\text{The proportion of } A_1 \text{ in the gametes} &= p^2 + p(1 - p)\\ &= p^2 + p - p^2\\ &= p\end{aligned}$$

This is mathematical proof that the proportion of an allele in the gametes will match its proportion in the population, in the absence of selection.

Terms and conditions

- The Hardy–Weinberg equilibrium is a balance in allele frequencies that can persist over generations when mating is entirely random; that is, when any gamete has an equal chance of combining with any other.
- This equilibrium will remain unchanged following meiosis and recombination. Because the effect of chance events is small, allele proportions are more likely to stay constant in large populations.
- The equilibrium will be maintained as long as there is no:
  i. selection,
  ii. migration,
  iii. mutation,
  iv. genetic drift.
- Migration and mutation are potential sources of new or additional alleles that will alter the frequencies within the population. Migration out of the population may also alter these proportions.
- Breaches of these conditions will lead to departure from equilibrium, but in a population large enough to undergo little genetic drift and with low rates of mutation and migration, selection may be detected as departures from the Hardy–Weinberg equilibrium (Box 3.2).
- Then a persistent change in allele proportions over several generations would indicate directional selection. A consistent rate of change—the increase or decline in the frequency of a particular allele—would suggest a consistent selective pressure.

Population genetics cannot tell us how much genetic difference or how many generations are needed to create a new species. It can compare the pace of microevolution and the intensity of major selective pressures in natural populations.

Box 3.2 Key technique

Detecting selection

The Hardy–Weinberg equilibrium provides a test of whether a population is subject to selection. We can compare the observed changes in genotype frequencies with those predicted by the null hypothesis, measuring their goodness of fit using the statistical test of χ^2.

i. The allele frequencies are measured in a sample of individuals. For simplicity, we shall say there are two alleles in the proportions p and q.
ii. This will produce genotypes close to the following proportions:

$$p^2 + 2(pq) + q^2$$

These are the expected frequencies predicted by the Hardy–Weinberg equilibrium, with no selection.

iii. From these proportions, we calculate how many individuals with each genotype are expected in a sample from the population (expected proportion × sample size).

iv. Having taken a sample of the population, we can calculate the difference between the observed (*O*) number of individuals with that genotype and the expected number (*E*); that is, $(O - E)$.

v. This is squared to get rid of negative numbers and then divided by its expected number: $(O - E)^2/E$.

vi. The results for each genotype are added together to derive χ^2:

$$\chi^2 = \Sigma(O - E)^2/E$$

vii. This calculated value is compared to tables of χ^2 (with $k - m - 1$ degrees of freedom (df) at 5% probability; where k = number of genotypes and m = estimated number of independent allele frequencies. Here, df = 1).

If the calculated value is greater than the tabulated value then the observed data are not in Hardy–Weinberg equilibrium.

Repeated departures from equilibrium over several generations would suggest the population did not meet the conditions for a Hardy–Weinberg equilibrium to become established, one possible cause of which is selection.

Coefficients of selection

Under a selective pressure different genotypes may have different reproductive successes and their proportions change. We can compare their fitness with a **coefficient of selection** (*s*) for each, relative to the fittest genotype.

We compare the frequency of a genotype before and after a pressure is applied, and measure the *change* caused by the selection. This is the genotype's fitness (W). The fittest genotype has the greatest increase, or the smallest decrease, and is set to 1. The selection coefficient for the other genotypes is their fitness relative to this genotype.

For example, the genotype A_1A_1 has the highest fitness (denoted W_0) so its s is set to 1:

$$A_1A_1 \; s = W_0/W_0 = 1$$

The changes in the frequencies of other genotypes are expressed as proportions of this:

A_1A_2 has a relative fitness of W_1/W_0 … and an s of $1 - W_1/W_0$

A_2A_2 has a relative fitness of W_2/W_0 … and an s of $1 - W_2/W_0$

Relative to A_1A_1, the other genotypes all have an *s* of less than 1. The larger the value of *s* the greater the selection against that genotype. For instance, an *s* of 0.2 would indicate that the survival of this genotype is 20% less than that of the fittest genotype.

- Selection coefficients can be used to predict the proportions of genotypes in subsequent generations.
- In artificial selection experiments, the selection pressure we apply is called the **selection differential (*S*)**. For example, if we select for breeding only those tomato plants which produce 10 fruit per truss rather than the population average of six, *S* is 4.
- In the next generation, we can measure the **selection response (*R*)**: this is the average difference in a trait between the offspring of those selected to breed and that of the original population. If our selected plants, when bred together, produce three more tomatoes per truss than offspring from the original population, *R* is 3.
- Because *R* is less than the applied differential, *S*, we know that the number of fruit produced per truss is not determined entirely by the genotype. Some environmental factor(s) must also determine this phenotype.

Evolution in the absence of selection

Few populations will consistently meet the terms required to establish a Hardy–Weinberg equilibrium. Departures from these conditions can lead to changes in allele frequencies, changes that we need to distinguish from the effects of selection.

Migration

Persistent gene flow into an isolated population from a larger population with different allele frequencies can prevent a local equilibrium being established.

- Consider an isolated population that has an allele (A_1) fixed. It receives a small number of migrants from a larger population also fixed at this locus, but these individuals all carry A_2. Before the migration the frequency of A_1 was 1.0; after the arrival of the migrants, it is less than 1.0...say:

  $A_1A_1 = 0.9$

so the migrants represent $A_2A_2 = 0.1$.

- Before mating occurs there are no heterozygotes, so the population does not have the frequency of genotypes predicted by Hardy–Weinberg. These would be:

  $$p^2 + 2(pq) + q^2 = 0.9^2 + 2(0.9 \times 0.1) - 0.1^2 = 0.81 + 0.18 + 0.01$$

- Because there are no heterozygous genotypes at this stage the adults are said to show a heterozygote deficit. Of course, heterozygotes will appear after mating and an equilibrium in allele frequencies will then be established, as long as there are no further migrations.

- With continued migrations, however, no equilibrium can develop and the heterozygote deficit persists, to varying degrees. The continued departure from the predicted frequencies is not a result of selection, but of migration.
- In this case local ecotypes may not develop due to dilution by gene flow from the parent population. An allele will need to confer considerable advantage to maintain its proportion in a local population under these conditions.

Inbreeding

Mating becomes increasingly non-random as population size decreases and the chance of mating with a genetic relative increases. Inbreeding may be detected from the frequencies of the different phenotypes, again departing from equilibrium predicted from the allelic proportions.

Adults fertilizing their own eggs represent the most extreme case.

- With self-fertilization all homozygous adults produce homozygous offspring.
- However, half of the offspring from heterozygous adults will be homozygous (A_1A_1 and A_2A_2) and half heterozygous:

		Egg	
		A_1	A_2
Sperm	A_1	A_1A_1	A_2A_1
	A_2	A_1A_2	A_2A_2

- With each successive generation, the proportion of heterozygotes is thus halved.
- No Hardy–Weinberg equilibrium will be found even though the frequencies of A_1 and A_2 may remain the same. It is the proportions of genotypes of zygotes (and adults) that change.
- With any significant inbreeding, we again observe a heterozygote deficit: a smaller proportion than that predicted by Hardy–Weinberg.
- The greater the degree of inbreeding in a population, the greater the heterozygote deficit.
- When only self-fertilization is possible, all heterozygosity will eventually be lost from a population.

With no change in allele frequencies there has been no evolution, but the low proportion of heterozygotes can have important evolutionary consequences for the population, most especially through **inbreeding depression**.

- This is the lowered fitness of individuals due to homozygosity at loci under selective pressure, most especially with recessive and deleterious alleles.
- In the heterozygous condition, a recessive allele may have little adaptive significance because its effects are masked by the other allele. However, some homozygous conditions can be lethal and inbreeding increases their frequency in the population.
- In other cases, homozygotes fail to compete against heterozygous individuals. The reasons differ between species and alleles, but examples include decreased growth rates or resistance to disease.

- The value of high heterozygosity explains the strategies adopted by plants and animals to avoid self-fertilization or breeding with close relatives.

  ➔ *Section 4.2, Life history and reproductive strategies*

Genetic drift

Isolating a small number of individuals, perhaps as colonizers of a remote habitat, effectively samples the genotype of the parent population. Consequently the new population may show a founder effect, with allele frequencies different from the parent population. Thereafter the small population can undergo genetic drift as its gene pool is sampled with each reproductive event.

- This persistent sampling effect over generations can cause genetic drift in any isolated population, but is more pronounced in small populations. Their genomes will differentiate more quickly from the parent population.
- A change in allele frequencies means the population has evolved. However, the relative success of an allele at each event is down to chance, not a selective pressure.
- In this case the change is not adaptive and is unlikely to increase the overall fitness of the new population: the differing phenotypic frequencies do not correlate with variation in reproductive success.

This form of evolution may be important in explaining rates of change in proteins and DNA sequences, many of which appear to be adaptively neutral.

- In the absence of migration or mutation, these proportions continue to change with each reproductive cycle. They will change rapidly if the zygotes represent a small proportion of the gametes available; that is, the sample remains small.
- The effects of sampling are cumulative, so that allele proportions in the gametes reflect the previous sample: an allele in decline is less likely to be sampled next time around. With sufficient time, genetic drift can cause any allele to be lost from the population.
- Equally, repeated sampling can lead to an allele becoming fixed: the only allele found at that locus. The probability of fixation through genetic drift (and in the absence of any selective pressure) is equal to an allele's initial frequency.
- As alleles become fixed or lost, heterozygosity declines. The more unequal the proportions of the alleles, the more homozygotes and the fewer heterozygotes result.
- The variations in allele frequencies will be at their largest in the smallest samples, but as sample size increases, more closely representing the gamete gene pool, a Hardy–Weinberg equilibrium will be approached.
- Genetic drift and loss of heterozygosity are less likely when reproduction takes large samples of a large population.

Box 3.3 **Looking for extra marks?**

The neutral theory of molecular evolution

Once we began to measure the variation in protein composition or homologous DNA sequences, we had to explain why rates of change at the molecular level were so fast. The rapid and regular fixation rates of alleles were not consistent with the view that most mutations were deleterious. Perhaps molecular evolution—change in allele frequencies and the proteins they coded for—was not driven by natural selection.

Kimura suggested that most molecular changes were selectively neutral, and fixation followed from the sampling effect of genetic drift. In his neutral theory, he asserted that beneficial mutations were so rare selection could not account for this speed or regularity and argued that the rate of molecular evolution was equal to the neutral mutation rate (nucleotide changes with no adaptive significance).

Evidence from DNA sequences which do not code for proteins (and are therefore likely to be selectively neutral) suggest the theory is largely correct: the fixation of mutated genes in these parts of the genome can be explained by drift. Elsewhere, changes in crucial DNA sequences are highly conserved: selective pressures mean that few changes persist over generations. The rates of fixation of silent mutations (those that do not change the amino acid sequence of a protein, or its functionality) also match those predicted by drift. Additionally, the regularity of the changes, both in the composition of proteins or of nucleotide substitution in functionless DNA sequences, reflect the probabilities of random sampling.

To demonstrate selection, the differences between species, populations, or individuals have to be shown to be adaptive or maladaptive. Since fixation by drift follows from a random process, the neutral theory can be used as a null hypothesis—rather like the Hardy–Weinberg equilibrium—to test whether molecular evolution has occurred at a faster rate than that predicted by chance.

Looking for extra marks?

You may need examples of variation in the trait that can be demonstrated to have no adaptive significance for an organism. Many of these refer to the 'junk' DNA in the genome or variations in the amino acid composition of some proteins. You will need to be able to review this as evidence for or against the neutral theory of molecular evolution.

Heritability

One way of assessing whether traits confer selective advantage is to measure their heritability, or the extent to which parental traits are seen in the offspring. Traits that show a high heritability are likely to have two properties:

i. they are genotypic adaptations with little phenotypic variation induced by the habitat;
ii. their code is dominant and conserved, and likely to have adaptive significance.

We expect to see variation between individuals, some of which is attributable to their genetic code and some to their environment.

We distinguish between:

- **quantitative traits**: those showing continuous variation, where phenotypes can show a range of values;
- **qualitative traits**, which occur as discrete phenotypes (often presence or absence).

Continuous variation is often associated with polygenic inheritance, traits coded by several genes, and also where the phenotype varies with environmental factors. Height in human beings is one example.

The inheritance of a trait is most easily decided with discrete variation.

➔ *Section 2.1, Genotypic and phenotypic variation*

The expression of a gene in the phenotype will reflect its interaction with the rest of the genotype and with the external environment: we may grow tall because we have the propensity coded over several loci, but have also acquired the resources to support that growth at crucial times in our development.

Heritability is the fraction of variability in a quantitative trait attributable to genetic variation. To measure heritability we compare the expression of the trait in the offspring with that of the parents: the greater the difference, the lower the heritability and the greater the environmental influence on the phenotype (Box 3.4).

- Most phenotypic traits are known to vary with both genotypic variation and phenotypic variation: with the inherited genotype and with acquired differences.
- The total variation observed between phenotypes in the population is denoted V_P. This arises from differences between individuals in their genotypes (V_G) and in their environments (V_E).
- Simple measures assume that genetic and environmental variation can be added together: $V_P = V_G + V_E$. This may not be the case if the expression of a gene depends in some way on its environment: we then need to include an interaction term as a further source of variation: $V_P = V_G + V_E + V_{G \times E}$.
- The gene or genes coding for a trait may also interact with other genes at different loci (epistatic interactions). Additionally, one allele may dominate another at the locus of interest. Together these interaction effects are called dominance variation (V_D).
- We can thus partition genotypic variation, V_G, into V_A and V_D: the simple additive variation at all the loci coding for the trait (V_A) plus the variation from interaction effects between genes (V_D).
- If we can assume that $V_{G \times E}$ (gene × environment) interactions are negligible, total phenotypic variation is then composed of

$$V_P = V_A + V_D + V_E$$

- Heritability (H^2) is simply V_G/V_P; that is, genotypic variation as a proportion of total phenotypic variation. This is termed broad heritability because it includes genetic interactions ($V_G = V_A + V_D$).
- Narrow or additive heritability (h^2) measures the effect of simple genetic (allelic) variation on phenotypic variation, without any interaction effects; that is, V_A/V_P. This simplification is assumed in many heritability studies (Box 3.4).

If we know the heritability of a trait (h^2), and apply a known selection differential (S), we can predict the selection response (R):

$$R = h^2S$$

This estimates how much a trait will change from one generation to the next, given its heritability and the selective pressure.

- Rearranging the equation, we can estimate heritability if we have a fixed S and have measured R:

$$h^2 = R/S$$

- A more reliable estimate of heritability can be made over several generations by plotting the cumulative selection response against the cumulative selection differential:

$$h^2 = \Sigma R / \Sigma S$$

The slope of this line is equivalent to additive heritability.

Finally, and to state the obvious, when there is no genetic element to the total phenotypic variation ($V_G = 0$) there can be no evolution. In this case, all of the variation is acquired and not inherited.

Looking for extra marks?

Have example data that compares heritabilities for different traits, preferably in one or two groups, with detail on the genotypes and the advantages they confer.

Box 3.4 Key technique

Measuring heritability

The fruiting performance of a tomato plant will depend on the genes it has inherited and the conditions under which it is grown. Taking several cuttings from a plant and growing them in different soils could tell us how important environmental factors are for the number of fruit per truss. If these clones all produce the same number it would suggest that truss size was almost entirely attributable to their genes, rather than the nutrient levels of their soils. Such experiments are most easily done for quantitative traits in species where we can control environmental factors and selection pressures precisely.

continued

The principle is simple: the closer offspring resemble their parents, the higher the heritability of that trait. We measure a trait's heritability by estimating the proportion of the variation in the offspring attributable to variation in their genotype:

$$\text{Heritability}\left(H^2\right) = V_G/V_P$$

where V_G = genetic variation and V_P = total phenotypic variation in the selected trait. V_P is the sum of V_G (the variance in genotypes between individuals) and V_E (the variance in their environment).

H^2 can range from 0 (entirely 'environmental') to 1.0 (entirely 'genetic'). This is called **broad heritability** because it incorporates all sources of genotypic variation, including the interactions between different genes and alleles (V_D).

However, for a direct correspondence of a genotype with a particular phenotype—having the gene and having the trait—we need to measure heritability without these interactions. This is the **additive heritability** (h^2), the variance in the trait attributable to the sum of variances at particular loci, with no dominance or interaction effects; the additive genetic variation (V_A) as a proportion of phenotypic variation:

$$\text{Heritability}\left(h^2\right) = V_A/V_P$$

This can be derived by plotting a quantitative trait in the offspring against an equivalent measure in the parents (Figure 3.2) and fitting a line using least squares regression. Least squares assumes additive variation, and the slope of the line is an estimate of h^2: offspring that perfectly match their parents in this trait would have a slope of 1. In this case the incidence of a trait in the parents guarantees its occurrence in the offspring, indicating the trait is entirely determined by the genotype.

This is the most commonly used measure in quantitative genetics, matching the incidence of a trait with the presence of one or more genes. Note that h^2 estimates the correspondence between genetic variation and total phenotypic variation under a particular set of conditions. We would get different estimates in different habitats with different genetic and environmental interactions. There is rarely a fixed heritability for trait, only an average derived from a range of conditions.

We thus need to measure this heritability over different habitats to properly estimate it for a population. Our experiments should ensure that offspring are not all grown under the same conditions as their parents, otherwise we would confound the environmental and genetic variation and be unable to distinguish the two.

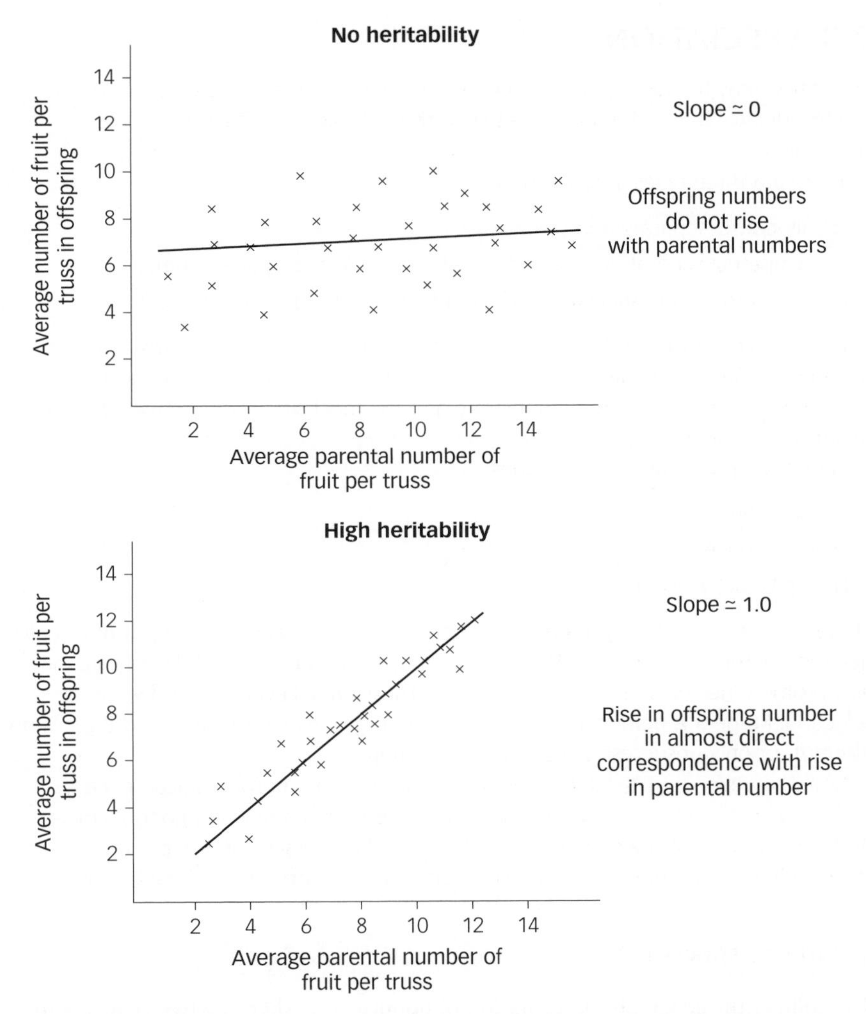

Figure 3.2 Heritability. We can measure the heritability of a trait as the slope of the line of best fit for a scatter of points that plot the average value for a trait in their offspring against that of the parental average. With perfect heritability in a population the slope will equal 1.0: a rise of one unit in the offspring corresponds to a rise of one unit in the parental average.

In this hypothetical example, the average number of tomatoes per truss (fruit cluster) for all offspring, over a season, is plotted against the average of the two parent plants. We need to repeat these measures for many tomato parents, over a range of conditions, to give a large data set encompassing the full range of this trait. If the slope is close to 0 there is little or no genetic component to the variation between offspring, and the differences are induced by the environment. The closer the slope gets to 1, the more variation we can attribute to their genes.

3.3 SPECIATION

Speciation may follow if gene flow ceases between populations because some form of reproductive barrier has arisen between them (Table 3.1). Their genotypes may then diverge.

There are three types of speciation:

i. allopatric: separation in space;
ii. sympatric: separation by habit or habitat, ecological space or time;
iii. parapatric: rapid speciation due to adaptation to local conditions.

There is no consensus on the validity of the three proposed mechanisms, although most evolutionary biologists accept that allopatric speciation—geographical speciation—is the easiest to demonstrate and the most significant, at least for the major groups of animals.

Speciation proceeds in three stages:

i. isolation,
ii. divergence,
iii. reproductive isolation.

Stage 1 can follow spatial, temporal, or behavioural separation. Stage 2 is principally genetic divergence, but physiological, anatomical, and behavioural divergence may also promote new semi-species or ecotypes. These are all evidence of directional selection. Stage 3 is demonstrated when the two groups are reunited and there is no attempt, or only unsuccessful attempts, at mating.

Note, however, that fertile hybrids can occur between morphological species, especially among plants. Reproductive isolation is more complete among animals, probably because of their complex developmental processes and the presence of sex chromosomes, although there are occasional examples of fertile animal hybrids.

Allopatric speciation

This follows the geographical separation of populations. There are two elements to this.

i. Intrinsic factors: the details of the organism's biology that cause a spatial separation. Species with a poor dispersal ability or a dependence on a particular habitat may be unable to cross even minor barriers.
ii. Extrinsic factors, principally the nature, scale, and persistence of a barrier to the movement of individuals. The larger the barrier and the longer it persists, the more likely an isolated population will become reproductively incompatible and *de facto* a new species.

Note that few barriers are impermeable to all species: the nature of the barrier will determine which species can cross it. A mountain range may be no barrier to some birds, yet some fish are readily isolated in adjacent but unconnected lakes. Both intrinsic and extrinsic factors are species-specific.

Allopatric speciation by colonization

Perhaps the most rapid speciation follows the colonization of a new habitat by a small group of individuals, after which migrations cease.

- The colonizers may start as divergent from the gene pool of the original population as a result of sampling (a founder effect).
- A small founding population will also undergo genetic drift. Drift does not adapt the new population to its niche, but it does promote a different genetic identity from the parent population.
- Drift can continue as long as there is no gene flow between the populations. For speciation, reproductive barriers need to establish and limit gene flow with the parent population (Table 3.1).
- Overall, drift is not thought to be a main driver of divergence and is insignificant compared to natural selection. Persistent and different selective pressures either side of the barrier will place the populations on distinct evolutionary trajectories.
- In a small population advantageous genes will spread rapidly. However, a founder population with limited variation at key loci will adapt slowly to the new conditions.
- With complete isolation, mutations are the only source of novel code for the new gene pool, and a small population has fewer opportunities for mutation—and few of these are likely to code for a change that improves the fitness of the phenotype.
- Should they arise, and confer some advantage, mutations can spread rapidly through a small population and perhaps become fixed in a few generations.

Allopatric speciation by division

A separation may not produce small populations, so there may be no founder effect and little risk of genetic drift. A physical division, perhaps by land movements, may divide a population into large subgroups.

- Separation by the development of a geomorphological feature can be rapid (a lava flow or landslip), slow (the creep of a glacier), or slower (the raising of a mountain range or the breaking up of a continent).
- The intrinsic factors which govern speciation include the species' generation time and net reproductive rate, as well as the speed and range of dispersal.
- Many of the phylogenetic divisions in the major groups of higher plants and animals can be linked to the creation of barriers and the break-up of continents. The spread and speciation of marsupials, after originating in Central Asia, is one example.

  ➔ *Section 8.3, Zoogeographical realms*

  Species separated by barriers can be reunited. Then the duration of isolation and the selective pressures in the two habitats will govern their genetic differentiation.

- With a short period of separation, mating between the groups may allow gene flow to resume and any genetic differences will decline as a single gene pool is formed.

Make the connection

Notice how speciation rates for different groups may be determined by the nature and speed of the environmental change.

This is an important interaction between a species and its environment, between the extrinsic and intrinsic factors that govern the speciation rates of different groups.

Consider why some groups—for example, some freshwater fish (such as sticklebacks)—speciate readily, whereas other fish (especially some sharks) are extremely ancient taxa. Some species become extinct with relatively minor changes in their habitat, others have adapted to changing conditions through geological time; again, the sharks are a good example of the latter.

Consider the possible connections with generation times, body size, and dispersal of eurytopic and stenotopic species. Good answers might contrast closely related taxa, and relate their differences to the permanence of their ecological space.

- The introduction of alleles from a disparate group or incipient species is called **introgression** and is known to be important in certain plant groups.
- A hybrid with a greater fitness than either of the parent groups will come to dominate the new gene pool.
- If hybrids do not survive, and gene flow remains limited, a reproductive barrier may be established.
- Such speciation is more likely if the parent groups are adapted to different ecological spaces, occupying different parts of an environmental gradient.
- The new species may co-exist without significant competition if they now occupy different niches. Otherwise competition will promote character displacement in one or both of the daughter species.

➔ *Section 2.2, Character displacement and speciation*

Sympatric speciation

This happens when neighbouring populations become adapted to different niches, without physical separation. This may follow character displacement and competition along one or more resource gradients.

Here reproductive isolation has to develop and be maintained, even though the two groups are living side by side. While not easily demonstrated in the wild, some of the best examples are found in fish in isolated lakes or rivers. The vast diversity of insect species of most ecological communities may also be attributable to sympatric speciation: their variety reflects the fine division of their niches and the increase in niche packing over time.

- Mayr suggested that sympatric speciation would be found in marginal habitats, where populations are isolated along some environmental gradient. Individuals of a herbivorous insect might switch to a different (but related) host plant from that exploited by the main population.

- Perhaps this plant occupies a different ecological time or space, and the colonizing insects become isolated from the parent population; shifting their activity to match the new host, they reproduce at a different time.
- Or they simply feed on a different part of the original plant. Because of the physical separation involved in these examples, some argue that this is a form of allopatric speciation.

However, sympatric speciation can also follow from disruptive selection and assortative mating. Disruptive selection promotes character displacement whereas assortative mating may allow pre-zygotic barriers to develop.

➔ *Section 2.5, Types of selection*

- Disruptive selection may produce forms able to exploit unoccupied parts of a resource spectrum.
- Assortative mating may be favoured because it produces fitter offspring and reinforces the character displacement resulting from the disruptive selection. Sexual selection then reinforces the new trait.

  ➔ *Section 2.5, Types of selection*

  - as each group becomes more specialized on its part of the resource spectrum the greater the advantage of the assortative mating for the offspring, and
  - the code for preferring a trait is passed on with the code for the trait itself: the more selective that assortative mating becomes, the more rapidly the character displacement proceeds.

These arguments suggest that sympatric speciation is driven by intraspecific competition. Schluter calls it competitive speciation, and suggests this as an explanation for rapid speciation of sticklebacks in the lakes of the Canadian Rockies.

Sympatric speciation also follows from the isolation of highly localized populations.

- Populations of the bent grass *Agrostis capillaris*, tolerant of high soil zinc concentrations, have increased self-fertility. Tolerant plants have flowering times different from those of non-tolerant plants. Both mechanisms limit gene flow between the varieties and promote genetic fixation in the ecotype so their offspring are homozygous for zinc tolerance.
- A change in the chromosome number can also lead to sympatric speciation. A substantial number of plant species have evolved after failing to preserve their chromosome number during meiosis.
- Others have crossed with closely related species, creating viable hybrids and adding to their chromosome number. Having more than two sets of chromosomes is termed **polyploidy**.
- Reproductive isolation may follow rapidly because successful fertilization (and cell division) is usually impossible between a polyploid and a normal gamete: polyploids may only mate and undergo cell division successfully with each other.
- Additionally, polyploidy will appear in a small number of individuals and this small group may then undergo genetic drift.

Speciation

Imbalances in chromosome number between the gametes causes major developmental problems for animals and this limits their hybridization. This is not the case for plants: most species do not have sex chromosomes—most are hermaphrodite—and their polyploids just grow bigger.

Many of our cultivated species, especially the cereal grasses, are the result of various forms of polyploidy and hybridization over the last 12 000 years.

Parapatric speciation

Some evolutionary biologists suggest that an additional type of speciation occurs in small, isolated, and largely immobile populations that undergo inbreeding and thereby adapt rapidly to the local conditions. These are ecotypes that have become highly localized species.

Again, polyploidy can create the same rapid genetic change in plants.

Revision tip

Ensure that you have examples of each type of (i) speciation and (ii) reproductive barrier. Ideally, use examples that connect these two elements, such as Hawai'ian *Drosophila*, or the cichlids of the East African Rift Valley lakes.

Check your understanding

Examination-type questions

1. Why is a high heritability likely to be indicative of some past selection pressure for a trait?
2. The table below shows the change in frequency of each genotype after a single generation in which the trait coded by the alleles was under selective pressure. Complete the table by calculating the fitness of each genotype and the selection coefficient it experienced. Ensure that you identify the fittest genotype and show your workings. What do you notice about the selection coefficient of the heterozygote?

Genotype	*Frequency in generation 0*	*Frequency in generation 1*	*Fitness W*	*Relative fitness*	*s*
A_1A_1	0.36	0.41			
A_1A_2	0.48	0.48			
A_2A_2	0.16	0.11			

You'll find answers to these questions—plus additional exercises and multiple-choice questions—in the Online Resource Centre accompanying this revision guide. Go to http://www.oxfordtextbooks.co.uk/orc/thrive or scan this image:

4 The ecology of the species

Key concepts

- The life history strategy of a species reflects the predictability of its habitat and availability of resources.
- Species make trade-offs between different selective pressures to maximize their reproductive success.
- Populations grow fastest when they occupy an ecological space close to their fundamental niche and when resources are not limiting.
- Life tables provide an age-specific summary of the schedule of deaths (and possibly births) for a population. With key factor analysis ecologists can identify the causes of mortality limiting total population size.
- The age distributions of populations differ according to their rate of growth: younger age classes dominate when there is rapid growth.
- Without migration, the rate of growth is given by the difference between births and deaths. When generations overlap this is measured as r, the intrinsic rate of natural increase.
- With few checks on population growth, natural selection favours individuals that maximize r.
- Such 'r-selected' species dominate habitats where resources are only periodically available because of their short life cycles and prolific reproduction.

continued

- Many populations are constrained by limited resources for which they have to compete. Their maximum number is the carrying capacity (K) for a population in that ecosystem.
- Competitive, 'K-selected' species tend to be long-lived, reproduce repeatedly, and have populations dominated by older individuals. Few species are readily categorized as perfectly r- or K-selected.
- Evolutionarily stable strategies are behaviours that dominate a population because most individuals adopt them.

Assumed knowledge

Basic mathematics.

Note that the models in this chapter run with either absolute numbers or population densities.

A conceptual map for this chapter is available on the companion website; go to http://www.oxfordtextbooks.co.uk/orc/thrive/ or scan this image:

4.1 AUTECOLOGY

We expect to find certain plants in certain locations, flowering at certain times of the year, growing for a certain length of time. The details of its natural history describe the plant's adaptations to its niche. Autecology is the ecology of the species: the fullest description of a species' ecological space, habits, and behaviour, informed by its evolutionary history.

- Autecology allows comparisons between closely related species in different niches and ecosystems, helping to identify the key selective pressures in these habitats (Figure 4.1).
- The similarities between related species in different niches reflect their shared phylogeny; for example, the behaviour of wolves and foxes in the tundra.
- Such studies can highlight the phylogenetic constraints on the adaptability of major taxa, with the physiologies and morphologies they have inherited. This inheritance explains, for example, the absence of liverworts in arid habitats.
- Comparisons between taxa also indicate the major selective pressures in different ecosystems; for example, the migratory behaviour of all large savanna herbivores.

Some species are fundamental to the structure or functioning of an ecosystem so that their autecology determines the nature of the habitat.

➔ *Section 6.3, Dominant and keystone species*

- A description of a species' niche includes its strategy to maximize its reproductive success (Box 4.1). This helps us to predict its population growth or identify the factors that limit it.

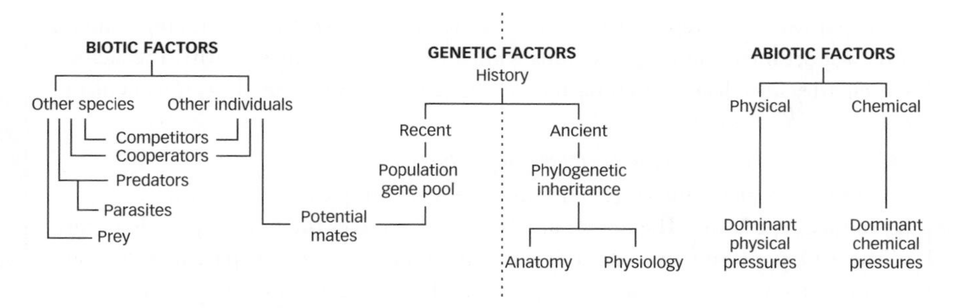

Figure 4.1 The range of factors which describe the ecology of an individual species. Autecology is rather more than the 'natural history' of a species, recognizing that a species is shaped by its evolutionary history as well as the current selective pressures defining its niche.

Factors to the left of the dividing line are those associated with the niche of a species and its specific adaptations.

Those to the right apply to it phylogenetic inheritance, the traits shared with related species living in the same or equivalent habitats.

- This is important for groups we wish to conserve or to harvest, or whose habitats we seek to protect.
- Most ecologists specialize in the autecology of particular groups, including their identification and classification. Such expertise is crucial to the practice of ecology.

Box 4.1 Make the connection

Trade-offs in life history strategies

Reproductive success depends upon two elements in the life of the individual: its reproductive output and its survival. A long life will mean more opportunities to reproduce, but this requires an investment in survival and securing the necessary resources. The alternative is to invest in reproduction, with a short life and a single, prolific, reproductive event.

Which strategy is best? Best is decided, of course, by the environment, and measured as the individual's reproductive success or fecundity. We can partition this into:

i. current **reproductive success**,
ii. future **reproductive value**.

Here, (i) is the reproductive output up to the current reproductive episode (the proportion of the next generation which are offspring of this individual) and (ii) is the individual's reproductive output in the future, during the remainder of its expected life. Natural selection operates on the sum of these two components—the **total reproductive success**—and the most successful are those most represented in the gene pool.

There are four basic reproductive strategies (see Table 4.1), according to the number of reproductive events and their timing. Species reproducing just once are

continued

semelparous, those reproducing several times are **iteroparous**; species reproducing after a short life cycle are precocious while others delay their maturity. Precocious semelparity and delayed iteroparity are opposite strategies, roughly equivalent to the *r*- and *K*-selected species.

➔ *Section 2.1, The eurytopic-stenotopic division*

Long-lived species grow larger and produce few young each season, but reproduce repeatedly. The longer an individual survives the longer its code is part of the gene pool, so survivorship itself adds to the fitness of the parent. However, reproductive success is usually *age-specific* and often declines with age. Some species are more prolific parents when they are larger, so the trade-off between early and delayed reproduction depends on the habitat and the autecology of the species.

Note the connections between the reproductive strategy and the constancy of the habitat, the availability of resources, and the size of the adult: the generation time, the number of offspring, the investment in parental care, the speed of growth, the speed of dispersal, and so on.

Note also that some plants and animals change their strategy in response to resource availability, especially in unpredictable habitats.

➔ *Section 1.2, Cost-benefit analysis*

4.2 LIFE HISTORY AND REPRODUCTIVE STRATEGIES

Some traits are easily quantified: the size of the adult, the age at first reproduction, the number of reproductive events, and so on. Consistent patterns of growth and reproduction are referred to as 'strategies' because they indicate a schedule for acquiring and deploying resources.

Life history strategies

A life history strategy determines when and how many resources should be devoted to growth or reproduction. The sequence of a species' life cycle in relation to ecological time is its **phenology**.

- Organisms adapt their growth and development to the periodicity of their habitat: the sequence, duration, and seasonality of resources, and the numbers and activity of their competitors and predators.
- Resources have to be allocated between different life stages. Many plants and animals have resistant stages that over-winter, or survive periods of drought or other adverse conditions.
- These resistant forms are often the main dispersive stage, such as seed or spores.
- Dispersal is important for genetic exchange between populations. Many higher animals use one gender to move between populations.

- Different stages may have different physiologies, morphologies, and behaviours to exploit different niches or habitats. In this way a species can avoid competition between age groups for the same ecological space.
- These schedules also optimize effort and coordinate reproductive activity with potential mates.

➔ *Section 2.2, Quantifying niche*

Revision tip

You should have one or more detailed examples of how a plant or animal allocates resources and schedules its reproductive activity for an ecosystem you have studied. Remember to include details of its dispersive stage(s).

Consider, for example, comparing the reproductive and resistant stages of an endoparasite, such as a fluke, with an ectoparasite, such as a flea, and relate their differences to the ecology represented by a mammalian host they share.

Reproductive strategies

Diverting most resources to reproduction may allow for early maturity, to support a single and prolific reproductive event.

- This has to be synchronized with other individuals, triggered by a reliable environmental signal.
- A short life cycle ending in a single reproductive event typically produces large numbers of gametes over a short period.
- This over-production of gametes allows for large losses, but only a few offspring need to become adult to maintain the population.
- The costs are the risks associated with a single reproductive event and these large losses.
- When conditions are favourable, a larger proportion survives. Parents able to increase their representation in the gene pool when times are good improve their inclusive fitness.

An alternative strategy is to delay reproduction, with resources diverted to parental growth.

- Investing in parental tissues improves the survival prospects of the parent, and perhaps its **fecundity**, allowing for multiple reproductive events.
- This strategy may result in fewer offspring at each event, but their survival rate will be higher with better provisioning of the gametes and, in some animals, with parental care.
- The other costs include the lost opportunities for early reproduction and the risk that an individual will not survive long enough to reproduce.
- The benefit is the spreading of risks over several seasons. Poor juvenile survival in one year does not mean the parent has lost all opportunities to reproduce.

Life history and reproductive strategies

The fitness of either strategy depends on the environment they inhabit (Table 4.1).

- The strategy can be distinguished by the numbers in the life cycle of an organism. Among others:
  - the number of reproductive events (Box 4.1),
  - the age at first reproduction,
  - the ratio of the sexes,
  - the size of eggs,
  - the average number of offspring in a brood,
  - the duration and cost of any parental care.
- Several of these are linked to adult body size. So, large organisms tend to:
  - reproduce later,
  - reproduce more than once,
  - make a larger investment in each egg and offspring.
- In each case, the opposite is true of smaller organisms.

Revision tip

Ensure you can explain how these measures are indicative of the two basic life history strategies described in the next section (*r*- and *K*-selected species).

Some plants and some animals have adaptable life cycles and may delay adulthood until the right conditions and resources are available. A strategy may also change with gender; for example, male angler fish remain small to become parasitic on a much larger female, effectively an attached male gonad. This is a useful adaptation in deep waters where partners are hard to find.

Reproduction			
	Early	vs	Late (precocity vs delayed)
	Single	vs	Multiple (semelparity vs iteroparity)
Offspring			
	Small	vs	Large
	Many	vs	Few
Parental care			
	Less	vs	More
Growth and development			
	Fast	vs	Slow
Adult size			
	Small	vs	Large

Table 4.1 Alternative strategies to maximize total reproductive value. The column on the left broadly corresponds to *r*-selected species; the right to *K*-selected species. Very often an attribute that increases reproductive success (such as early reproduction) has to be traded-off against another desirable attribute (multiple reproductive events). Some species are able to switch strategies when resources demand the alternative for greater reproductive success.

➔ *Table 2.1*

r- and *K*-selected species

Rapid population growth is only possible in organisms which themselves grow rapidly. Adults with small body sizes develop quickly and have short generation times. Species which follow this reproductive strategy are described as ***r*-selected**.

➔ *Table 2.1*

- Rapid reproduction is the only viable strategy when resources are available for short or intermittent periods, or in unpredictable habitats.
- This strategy requires individuals to disperse readily to exploit resources when and where they are available. Moving between habitats means they must also survive a variety of environmental conditions.
- Fast growth and development is possible in an unpredictable environment where there are few established competitors, or resources are abundant.
- *r*-selected populations grow and decline as resources wax and wane. Their numbers rarely achieve any constancy.

Large adults need time to acquire the resources to mature. Because they have a long generation time (from zygote to first reproduction) their population growth is relatively slow. Species having to compete for limited resources will become closely adapted to one ecological space. This strategy is described as ***K*-selected**.

- A delayed reproduction may be the only viable strategy when resources are limited, either by supply or by competition. This strategy will work if the ecological space persists long enough for reproduction (Table 4.2).
- Large-bodied organisms are more effective competitors. Occupying a large part of a resource constrains competitors.
- Crowding out the competition means the highest mortality is among the juveniles: tree seedlings grow slowly in a mature woodland and may not reach the canopy unless older trees are lost.

	Large	Small
Costs	High maintenance energy	Less effective homeostasis
	High growth and reproductive energy	Less effective competitor/predator
	More prone to resource shortages	Smaller offspring
	(More likely to be predated)	More likely to be predated
	Long generation time	
Benefits	More effective homeostasis	Low maintenance energy
	More effective competitor/predator	Less prone to resource shortages
	(Less likely to be predated)	Low growth and reproductive energy
	Larger offspring	Short generation time
	(More offspring)	(More offspring)

Table 4.2 The costs and benefits of being large or small as an adult. Some of these costs/benefits depend upon the habitat and the strategy adopted (shown in parentheses).

- Being large limits the mobility of some species. However, the ability to colonize new ecosystems may confer few advantages on an organism closely adapted to a highly predictable habitat.
- *K*-selected populations do not undergo rapid change. They are principally constrained by intraspecific competition and their population stays close to its maximum number for their habitat.

These alternative strategies are most obvious in plants: compare the growth and reproductive strategy of oak trees in a temperate woodland with poppies flowering in a ploughed field. The relative size of their seed is a good proxy for the size of the parent, the species' generation time, and their capacity to disperse.

- The seeds of many weed-like plants can germinate in a wide range of disturbed soils and can travel far. Acorns do not travel without the aid of animals, but do have the resources to support early growth in a crowded woodland.
- Large eggs require many resources and represent a large loss if they fail to become adult. In contrast, the loss of many small eggs is easily balanced by a few offspring surviving.
- A large investment in eggs and offspring may not be made every year and missed opportunities lower the reproductive value of a parent. Nevertheless, this is the trade-off a *K*-selected parent can make, between its current reproductive success and its future reproductive value (Box 4.1).

4.3 LIFE TABLES

Mortality rates differ between populations and between age groups within a population. The groups suffering the highest deaths are likely to determine the size of a population. In some cases experimental studies can identify the causes of this mortality and its capacity to limit growth. For most species we have to collect data in the field and construct a life table.

In its various forms, a life table is a summary of age-specific mortality rates. This is based on data from a number of censuses or by following a number of cohorts through their life cycle (Box 4.2).

- Life tables allow derivation of survivorship curves. These are a plot of the number alive (usually its logarithm) against age or alternatively the proportion surviving.
- This is effectively the average life expectancy of an individual of a particular age. Many species show characteristic survivorship curves which reflect their life history:
  - type 1: high mortality only in older age groups,
  - type 2: consistent rates of mortality throughout the life cycle,
  - type 3: high mortality in the younger age groups.
- More useful is a key factor analysis, where a series of life tables is used to identify the significant causes of mortality for each age group (Box 4.3). The 'key factor' is

Box 4.2 Key technique

Life tables

The life history strategy of a species depends on the age classes when mortality rates are highest. The principal method of summarizing age-specific mortality is to create a life table for a species.

There are two types: static (or stationary) and cohort tables. Static life tables are constructed from a census of the numbers alive in each age group, recorded on one occasion; a cohort table follows a group of individuals 'born' (recruited) at a specified time, and recording the number of deaths as the cohort ages, until it expires. Either method may count both sexes, or just females (according to the autecology of the species). The reliability of each improves with the number of surveys, and with large sample sizes.

We begin by deciding the most appropriate age interval for a species, given its life expectancy and stages of development. In a cohort life table, the initial sample size (n_0) and the number of survivors entering each subsequent age group (n_x) are needed. This allows us to calculate the proportion of the original cohort surviving to the end of each interval (l_x), the number dying during that interval (d_x), and the rate of mortality within that age group (q_x).

Thus the only data needed is the size of the remaining cohort as each age interval commences: if the cohort consisted of 1000 individuals at the outset ($x = 0$) and there are 600 alive at $x = 1$ (that is, $n_1 = 600$) then during this interval:

$$\text{Proportion of cohort surviving } l_x = 600/1000 = 0.6 \left(\text{that is, } l_x = n_x/n_0\right)$$
$$\text{Number dying } d_x = 1000 - 600 = 400 \left(d_x = n_{x-1} - n_x\right)$$
$$\text{Mortality rate } q_x = 400/1000 = 0.4 \left(q_x = d_x/n_{x-1}\right)$$

Continuing, at the end of the second age interval $n_2 = 450$, so:

$$l_x = 450/1000 = 0.45$$
$$d_x = 600 - 450 = 150$$
$$q_x = 150/600 = 0.25$$

Note that l_x is derived as a proportion of the original cohort size throughout.

Cohort life tables have to be used for species with discrete generations (usually annual life cycles). They may also be used with overlapping generations.

Static life tables assume static populations, and record the mortality rates only at the time of the census. It would only give the same result as a cohort life table in a constant environment, with a population close to its equilibrium size, and without migration. These conditions are rarely met because both the habitat and rates of births and deaths change over time. Cohort tables can work well in

continued

populations in which there is little or no migration, but require repeated measurements over the life of the cohort.

These tables give the age distribution for the population; that is, the proportion in each age category (Figure 4.2). They also help identify the age-specific mortality most likely to limit the size of population, using key factor analysis (Box 4.3). With tables of fertility for each age group of females, life tables are used to calculate mean reproductive values (Box 4.1) and age-specific fecundity.

The same calculations can be done with **population densities** rather than absolute numbers. For plant population studies, where individuals are not easily distinguished, biomass is often used as the most convenient measure.

the cause of mortality most closely correlated with total population mortality, and likely to be limiting population size.

- A life table may also be used to summarize the fecundity schedule for a species, and identify the age groups that produce most 'recruits' to a population.
- The conservation of commercial fishery stocks depends on such analyses. It is often the large and elderly that produce the most eggs, the fish preferentially caught by our trawler nets.
- The most complete life tables are for human census data and are used by insurance companies to calculate premiums.

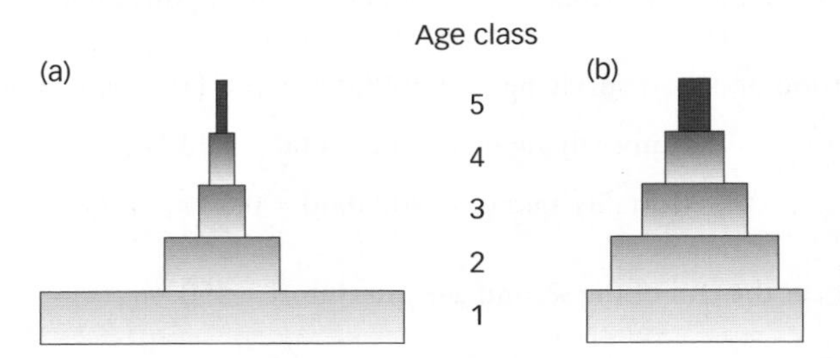

Figure 4.2 Two age distributions which remain unchanged over time.

In a stable age distribution (a) the population is growing exponentially and has fixed rates of births and deaths in each age group, so the proportions in each age class stay the same. Their rapid growth means the younger age classes dominate.

With a stationary age distribution (b) overall birth rates are balanced by death rates, so the population is not growing. Here the population is close to its carrying capacity. Under these conditions a more even distribution develops. Under other conditions the age distribution is not fixed.

Box 4.3 Key technique

Key factor analysis

Key factor analysis uses the data from a life table to partition the total mortality of a cohort between age groups and between different causes, to identify those factors most likely to determine population size.

The number lost from each age group is expressed as its *k* value. This is the difference between the log of the number entering this group (n_x) and the log of those joining the next age group (n_{x+1}): simply the mortality (or emigration) from the cohort during that interval (x to x_{+1}). As logarithms we can sum them to get the total mortality (K) and to compare mortalities between age groups.
For example (continuing from Box 4.2):

$$n_0 = 1000 \quad \log n_0 = 3.000,\ k_0 = 0.222\ldots \quad (\text{that is, } 3.000 - 2.778)$$
$$n_1 = 600 \quad \log n_1 = 2.778,\ k_1 = 0.176\ldots \quad (2.778 - 2.602)$$
$$n_2 = 400 \quad \log n_2 = 2.602,\ k_2 = \ldots \text{ and so on}$$
$$K = k_0 + k_1 + k_2 + \ldots$$

There is unlikely to be a single cause of death associated with each age group, so we have to partition the mortalities between several causes and give each a *k* value. For example, for age class 1, we might identify three major causes of death:

$n_{1a} = 600$ viral infection, n_1 dying 20 — $\log n_{1a} = 2.778,\ k_{1a} = 0.015$
$n_{1b} = 580$ predation, n_1 dying 140 — $\log n_{1b} = 2.763,\ k_{1b} = 0.120$
$n_{1c} = 440$ parasitism, n_1 dying 40 — $\log n_{1c} = 2.643$, $k_{1c} = 0.041$
($n_2 = 400$ — $\log n_2 = 2.602$)

$$k_1 = k_{1a} + k_{1b} + k_{1c}$$

The sum of these values will equal the *k* value for the n_1 age group (0.176). Predation is seen to be the largest single cause of mortality in this example.

(Here we are not working with sequential age groups. Instead, think of the partitioning as 'of the 600 in this age group, 20 die of a viral infection; of the 580 remaining, 140 die of predation; of the 440 remaining, 40 die of starvation'. Four hundred survive into the next age group.)

Records from several years allow us to decide which factors are most important in determining total population mortality: changes in the *k* value of a key factor will tend to correlate with changes in total mortality. Visual inspection of plots (Figure 4.3) gives an indication of the most likely key factor. Again, the analysis can use densities as well as absolute numbers.

We can also compare regression coefficients for each factor with total mortality. Those causes of mortality with coefficients close to 1.0 are likely to be the most influential on total mortality and the largest coefficient will indicate the key factor. Note that this may not be the factor causing the largest number of deaths: if it does not follow the changes in total mortality it cannot explain them (Figure 4.3).

In some species, and at certain ages, emigration may be the key factor for total population size. Identifying and counting emigrants will depend on the species' autecology and the methods available to the ecologist; in a cohort study, we only count the survivors from our identified cohort. With static life tables, we may also need to measure migration.

continued

Not only do logarithms allow the simple addition of key factors, they also allow simple comparisons between populations and between species. However, this assumption of additive mortality is not applicable to all species, especially those with overlapping generations.

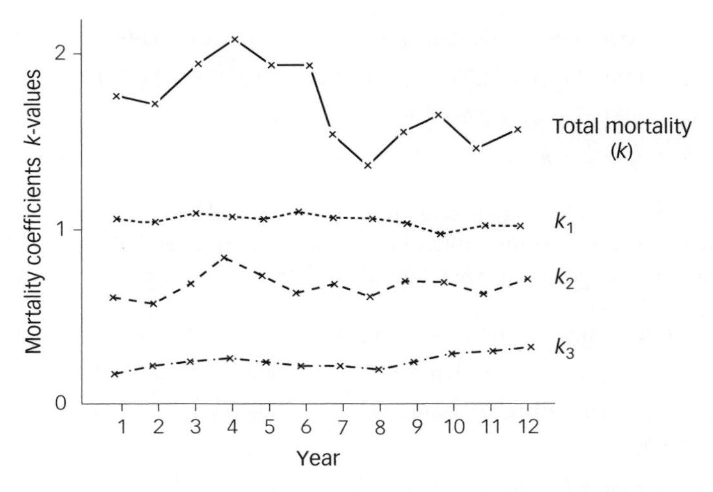

Figure 4.3 Determining the key factor controlling the size of a population. In this hypothetical example, three *k* values associated with three causes of mortality are plotted over several years, alongside the total population mortality. Note that k_1 is the factor that causes the largest loss of individuals, but is relatively consistent from one year to the next. k_2, on the other hand, has less effect, but shows a good match with total mortality. k_2 is likely to be the key factor.

Revision tip

Ensure you understand the calculations in a cohort life table. You should be able to construct a life table from raw data given numbers surviving at the end of each age interval.

Age distributions

The proportions of a population in different age groups determine its capacity for increase: a rapidly growing population is recruiting to its youngest age classes (through hatching, birth, or germination), some of which will later become productive.

In many species the probability of being a parent, or of dying, is age-specific, and, in animals, may also differ between the sexes. Knowing the fecundity and survival rates in different age classes allows us to predict the growth of a population.

- Cohort or static life tables from several years are used to establish the age distribution, and to measure its variation over time.

- Deciding the age limits for these classes depends on the species' autecology and the practicalities of making reliable counts in the field. Generally, the biology of the stage is more important than the elapsed time.
- Many animal studies count only females, since these alone can produce offspring.
- If numbers in each age class always increase at the same rate, the age structure of the population will remain the same: age-specific birth rates and death rates are unchanging so each class grows at a fixed rate and their relative sizes are fixed. This is termed a **stable age distribution** because its shape, the proportions in each age class, remains the same (Figure 4.2a).
- Since each age class maintains the difference between its birth and death rates, the overall population has a constant reproductive rate and it grows geometrically (see Section 4.4, on models of population growth). Such stable age distributions are 'bottom heavy', with most individuals in the youngest age classes.

  ➔ *Section 4.4, Exponential growth with a fixed reproductive rate*

An age distribution may also be unchanging when there is little population growth.

- This happens when the *overall* birth rate is matched by the *overall* death rate. This **stationary age distribution** is characterized by lower birth rates, and higher death rates, especially among the younger age classes.
- The age distribution is more uniform so that a larger proportion of individuals is found in the older age groups (Figure 4.2b).
- This is characteristic of populations close to their maximum size, where the environment is relatively constant and resources are limited.

Few populations achieve a stationary age distribution because birth and death rates fluctuate with environmental conditions. Similarly, stable age distributions are relatively rare because few populations can continue to grow without limit.

Looking for extra marks?

Note that a stable age distribution over several generations or different populations is evidence of an *r*-selected species.

Similarly, a stationary distribution would be indicative of *K*-selected species.

4.4 MODELS OF POPULATION GROWTH

A population is a collection of interbreeding individuals alive in a particular place at a particular time. Definitions of time and place can be problematic if individuals readily cross our notional boundaries. What counts as an individual can also be problematic, especially among plants and those animals which show modular or clonal growth.

The growth models presented in this section are for single species, and assume discrete populations with no migration. They assume individuals we can count, or estimate as densities (numbers per unit area or volume).

Population growth in unlimited environments

This describes growth when there are no checks on the population.

The simplest models are for populations that live for just one year—many plants, insects, fish, and amphibians are annual species—emerging, reproducing, and dying within a single season. These have non-overlapping generations and survive between seasons using a dormant stage: seeds, spores, eggs, or larvae.

Non-overlapping generations

- The number alive in the next generation depends on the number alive in this generation. Counting the adults at maturity each year, we can work out the multiplier between one generation and the next. This is the population's **reproductive rate, R_o**:

$$R_o = N_g / N_0$$

- N_0 = number adults at start; N_g = number adults in the next generation.
- If the population doubles, $R_o = 2.0$. This means that, *on average*, each individual leaves two offspring.
- If R_o is fixed this produces geometric population growth. From this it is easy to predict the size of the population in subsequent generations:

$$N_1 = R_o N_0$$
$$N_2 = R_o N_1 = R_o\left(R_o N_0\right) = R_o^{\,2} N_0$$
$$N_3 = R_o N_2 = R_o(R_o^{\,2} N_0) = R_o^{\,3} N_0$$
$$\text{so } N_t = R_o^{\,t} N_0$$

You can see this plotted in Figure 4.4.

Overlapping generations

When generations overlap, some adults survive for more than one season. Individuals die at different ages.

A census of the adult age class, taken at the same time each year, will incorporate both the births and deaths since the previous count. These counts no longer represent one cohort but several: the size of the population (N) is now measured after a known time interval (t):

This change is termed the **net reproductive rate**, R_N:

$$R_N = N_t / N_0$$

N_0 = number adults at start; N_t = number adults after time t.

- R_N is the *average* rate of change per individual per unit time, allowing for births and deaths.

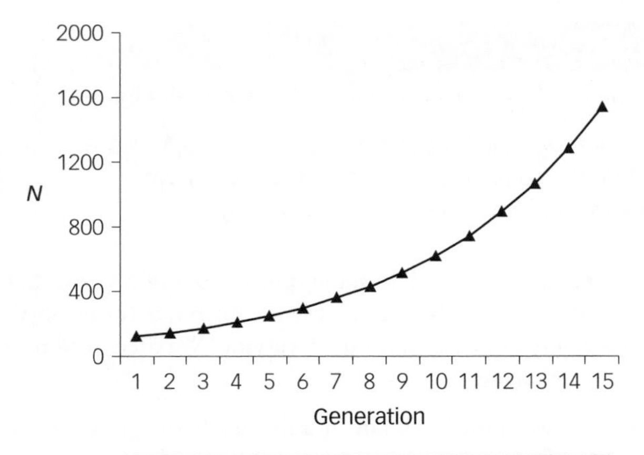

Generation number	*N*	Increase in *N*
1	120	20
2	144	24
3	173	29
4	207	35
5	249	41
6	299	50
7	358	60
8	430	72
9	516	86
10	619	103
11	743	124
12	892	149
13	1070	178
14	1284	214
15	1541	257

Figure 4.4 Population growth in an unlimited environment, for an organism with discrete generations. This has a starting population of 100 and a constant reproductive rate of 1.2, rounded up to give whole numbers.

- The population grows when $N_t > N_0$ and consequently $R_N > 1$. If deaths exceed births, $N_t < N_0$ and $R_N < 1$.
- A simple plot of N with a fixed R_N is given in Figure 4.4, as a series of discrete steps. Each N_t is the previous population size (N_{t+1}) multiplied by R_N. The increase in N is the difference between N_t and N_{t+1}.
- The characteristic shape of such a plot is a j-shaped curve: the exponential growth of a population in an unlimited environment. Any value of R_N above 1.0 will produce this curve: higher values create steeper curves.
- Values of R_N below 1.0 imply that deaths outstrip births and the population is in decline.

Note that R_o and R_N are only equivalent when $g = t$ (when the sampling interval is one generation). Both then measure the *average* change per individual over a generation.

Revision tip

Ensure that you can follow the calculations in each of the figures.

These models are available for you to explore their properties in Worksheet 2 on the companion website go to http://www.oxfordtextbooks.co.uk/orc/thrive/ or scan this image:

Deaths may occur at any time and, in some species, so might births. Such populations show continuous change because the birth rate (or natality rate) and death rate (or mortality rate) are continually varying. We then need to use instantaneous rates of change.

- The difference between the birth rate and the death rate gives the average rate of change per individual:

  $r = b - m$

  r, the instantaneous rate of change per individual, is the difference between the instantaneous birth rate per individual (*b*) minus the instantaneous mortality rate per individual (*m*).

- *r* is the intrinsic rate of natural increase. Instantaneous rates are expressed as logarithms to allow us to subtract mortality rates from birth rates and derive *r*. So *r* is equivalent to the logarithm of R_N (by convention natural logarithms, log_e or *ln*, are used).
- The change in the population size (d*N*) over the time interval (d*t*) is simply the current population size multiplied by *r*:

  $dN/dt = rN$

- This differs from the previous model only by calculating the change over a very small interval of time (d*t*).
- With a fixed value of *r*, the population grows as previously described: exponential growth with the speed of growth set by the size of *r*.
- When $b - m = 0$ there is no change in the population. A population will grow if $r > 0$, and decline if $r < 0$.
- R_N is a 'finite' rate because it measures the rate of change over a discrete interval of time. It is easy to switch between finite and instantaneous rates of change given the relation:

  $R_N = e^{rt}$

  where *e* is the base of the natural logarithm.

- Since $r = \log_e (R_N)$ (taking *t* to be 1.0) we can re-write this as:

  $\log_e\left(N_t/N_0\right) = r$

- So if we have two population counts separated by the time interval *t*, we can calculate *r*.

If, over one year ($t = 1$) the population doubles:

$$\log_e(200/100) = 0.693$$

That is, an R_N of 2 represents an instantaneous population growth rate (r) of 0.693 per individual per year.

Note that these models work in the same way when dealing with population densities.

Looking for extra marks?

Deriving *r*

r is the instantaneous rate of change per individual, the balance of instantaneous birth and death rates, and so will equal the natural logarithm of the net reproductive rate R_N:

$$r = \log_e(R_N)$$

Since $R_o = R_N$ when the interval between population counts is one generation then

$$r = \frac{\log_e R_o}{T}$$

where T is the average length of time for a generation.

We can thus estimate r from a cohort life table by calculating R_o and working out the time to complete one generation.

Maximum rate of population growth

Natality and mortality rates define the life history strategy of a species. Slow-growing species have a low intrinsic capacity for population growth ($\mathbf{r_{max}}$, the maximum rate at which the population could grow under ideal conditions). This follows from their long generation cycles, their small number of offspring, and the extended time over which they produce them.

In contrast, habitats that favour rapid population growth have species with high r_{max} values (typically *r*-selected species).

Make the connection

r_{max}

The full reproductive potential of a species will only be achieved when there are no checks to slow its time to maturity, or limit the resources it can devote to reproduction.

That few populations come close to reproducing at their r_{max} is an indication of the selective pressures they are under. We can, from a study of their reproductive physiology and autecology, estimate a species' reproductive potential, and thus the extent to which they are limited by their habitat.

continued

> *Note that populations some distance from their fundamental niche have to make trade-offs and adjust their allocation strategies to maximize their total reproductive success. Consider the range of factors that may cause resources to be diverted from reproduction when living in marginal habitats.*

Population growth in a limited environment

No species can sustain unlimited growth over the long term: a lack of resources or a build-up of waste limit the numbers a habitat can support. Collectively these represent the 'environmental resistance' to further population growth.

- The **carrying capacity** (K) is the maximum population size that a habitat can support. This is set by the resource 'space': as numbers increase, so does the intensity of intraspecific competition.
- If other species exploit the limiting resource, interspecific competition will reduce this capacity. The models in this chapter consider only intraspecific competition.
  ➔ *Section 2.2, Competition; Section 5.2, Models of interspecific competition*
- Some species never reach their carrying capacity because their numbers are checked by predation or parasitism. For these populations competition for resources may never limit N.
- Assuming that intraspecific competition sets the carrying capacity for a species, we can incorporate this into a model by calculating the unused capacity for population growth in a habitat:

  $$\frac{K-N}{K}$$

  where N is the current population size and K is the maximum number the ecosystem can support.

If the total capacity is 500, and there are 100 currently occupying the resource space then:

$$500 - 100/500 = 0.8 \text{ of the capacity is available } (\text{or } 400 \text{ unoccupied spaces})$$

- The intensity of this competition, and thus environmental resistance, increases as the availability of the resource declines.
- Incorporating this into the previous model checks the growth of the population:

  $$\frac{dN}{dt} = rN\frac{(K-N)}{K}$$

  $K - N/K$ is an adjustment applied to r according to the unused capacity: as the available resource declines, so the rate of population increase slows. When K is fully occupied, growth stops (Figure 4.5a). The result is logistic population growth.
- This makes biological sense: a shortage of resources supports less reproduction (b falls) and less growth (m rises). When $b = m$, $r = 0$.

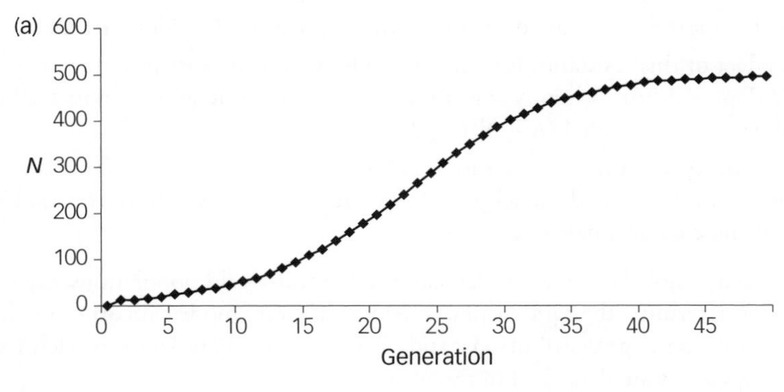

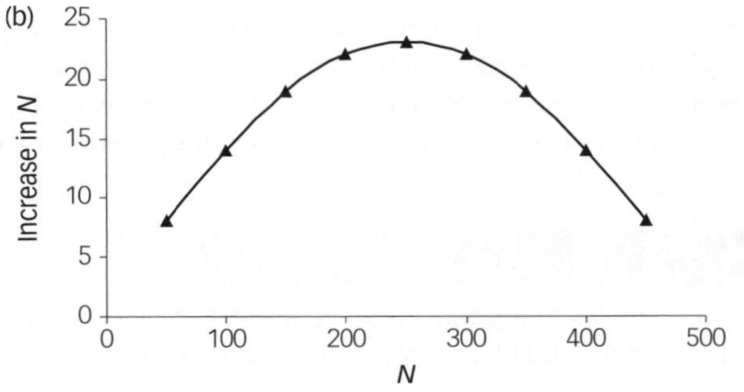

$K = 500$	$r = 0.18$		
N	$K - N/K$	$r \times (K - N/K)$	$dN/dt = rN(K - N/K)$
50	0.9	0.162	8
100	0.8	0.144	14
150	0.7	0.126	19
200	0.6	0.108	22
250	0.5	0.090	23
300	0.4	0.072	22
350	0.3	0.054	19
400	0.2	0.036	14
450	0.1	0.018	8
500	0	0	0

Figure 4.5 Population growing in a limited environment.

(a) Here we model overlapping generations, with the carrying capacity (K) set to 500. The rate of increase per individual in the population (r) is 0.18.

(b) The effect of environmental resistance is shown by calculating the size of increase at different population sizes. For each N, the environmental resistance is calculated as the proportion of the capacity still available (K – N/K):

$$\text{for } N = 50 \text{ this is nine-tenths } \left(500 - 50/500 = 0.90\right)$$

Multiplying this by 0.18 *(r)* gives the actual rate of increase:

$$(0.18 \times 0.90 = 0.162)$$

Multiplying the result by N gives the net increase at that population size:

$$(0.162 \times 50 = 8)$$

Notice that the largest increment is at $N = 250$, half of the carrying capacity. The increase in N declines as the habitat fills, so that no increase is possible at K.

- Thus the reproductive rate decreases as the population density increases.
- The effect of this resistance is seen with a plot of the increase in N for each value of N (Figure 4.5b). When $N = K$ no increase in the population is possible because all the space is occupied ($K - N/K = 0$).
- Notice the greatest rate of increase is at half the carrying capacity. This is where both N and the unused capacity are still large. Moving away from this midpoint, one of these determinants decreases.

Despite their simplicity, these models can predict real-world populations, especially if few factors determine their growth rates. More elaborate models incorporate the effects of time lags, age distributions, and chance factors. Population models for two interacting species are described in the next chapter.

The intensity of intraspecific competition will favour individuals able to exploit the limited resource efficiently. 'K-selected' species have traits that support their competitive ability rather than rapid population growth. These models work best for such populations, remaining close to their carrying capacity and not showing major fluctuations.

Revision tip

You are likely to need examples to support these models. You may need to know r_{max} values for different groups and be able to relate these to a species' life history strategy and habitat.

4.5 POPULATION VARIABILITY

Key factor analysis and other studies suggest a range of factors determine population size and constancy: for most species resource availability is just one variable in a multivariate world.

- When a population ceases to grow, birth rates equal death rates. This will occur close to K (Figure 4.6a). When $b < m$ the population declines.
- Both b and m may change with the size or density of the population. They are then described as **density-dependent**. This implies a feedback, with rates changing as population density changes. In Figure 4.6a, the death rate increases and the birth rate falls as the density increases, so providing negative feedback.
- Factors which reduce b or increase m as density increases can each regulate a population, either together or alone. With only minor changes in their habitat, these populations maintain a consistent size over the long term.
- When *both* b and m are density-independent, there is no feedback and the population is likely to be highly variable.
- It is then possible for both b to *increase* and m to *fall* as population density rises. In either case the population continues to grow so no equilibrium density is achieved.
- Natality and mortality rates change as the habitat changes. An increase in b, if m remains constant (Figure 4.6b), allows for a larger equilibrium density and implies an increase in the carrying capacity.

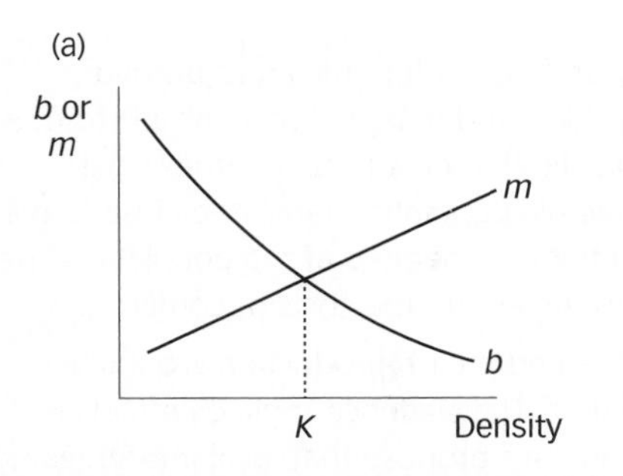

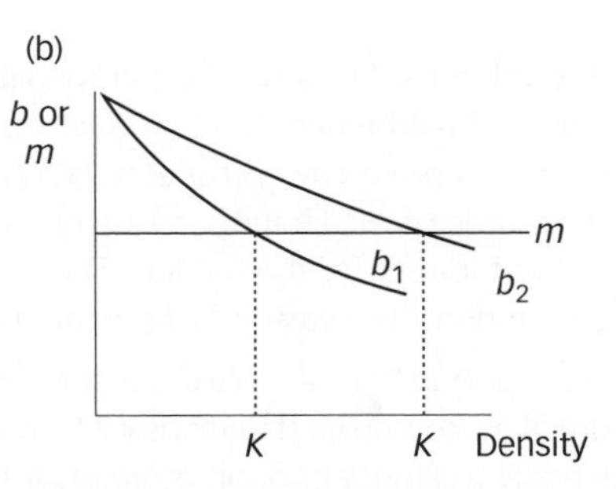

Figure 4.6 (a) A population reaches a stable equilibrium density (K) when $b = m$. Both b and m are density-dependent.
(b) Here mortality is density-independent and b is density-dependent but stable equilibria are still achieved. A slower rate of decline in b (b_2 compared to b_1) with a fixed m allows for a higher K.

- Rates also differ between populations. Some populations may use a resource more or less efficiently or occupy a smaller or larger range.
- Additionally, the effect of density on these rates is not always linear; for example, b may be depressed at very low population densities because encounters between mates are infrequent. Such effects are crucial when trying to conserve endangered species in the wild.
- Some suggest that interspecific competition can result in a form of group selection causing a population to regulate itself close to its carrying capacity, by allocating resources between individuals. However, the evidence for this is highly contested.

Make the connection

r- and K-selection

Note the several places where aspects of this classification are used to explain a species' adaptations to the variability of its habitat.

➔ *Section 1.2; Section 2.1; Table 2.1; Sections 4.2, 4.3; and Tables 4.1, 4.2*

Ensure that you can explain why stable and stationary age distributions are expected in habitats with abundant or limited resources respectively. Consider how, in turn, these explain the life history strategies of the two types, illustrated with examples.

Box 4.4 — Looking for extra marks?

Group selection and population regulation

Group selection was proposed by Vincent Wynne-Edwards as a mechanism of population regulation. He suggested that some animal populations avoided depletion of scarce resources, and a population crash, by individuals limiting their reproductive efforts when they detected shortages. Various mechanisms were proposed, most famously the territoriality of lions, with prime areas divided between the dominant (and productive) prides, and marginal habitats left to those with poor reproductive prospects. Not only did the most fit secure the largest (lion's) share of the limiting resource, the whole population remained close to the carrying capacity for the habitat. This partitioning operated at the population level, avoiding both the waste of unallocated resources and the costs of conflict.

Thus, a population of individuals that restrained their reproduction would be more likely to persist than those which didn't. The evidence for such effects is generally lacking, although more recent versions propose that local interbreeding groups, or **demes**, might enjoy higher fitness from some shared traits in a form of group selection. Nevertheless, this is some distance away from population regulation.

Not all density-dependent rates have an effect on population size. Many factors may be density-dependent, and have some effect on b or m, but may not be key because:

- their effect is too small to check population growth;
- other factors may be more important in controlling total mortality: these may not be density-dependent;
- with a population responding to several factors, changes in one can lower or increase the selective pressures of another, so the response is often a trade-off;
- migration plays an important role in many population's dynamics.

Plotting mortality factors (k values) against age-specific population density indicates which are density-dependent. A key factor will explain the major shifts in total mortality and population size (Figure 4.7).

These models assume that each individual is equal and their birth or death is equivalent to any other. This is often not the case: the death of an individual that has ceased reproducing has different implications from that of an individual yet to reproduce. b and m are average rates and are appropriate for simple models, but preferential predation of some age groups, for example, can alter b as well as m.

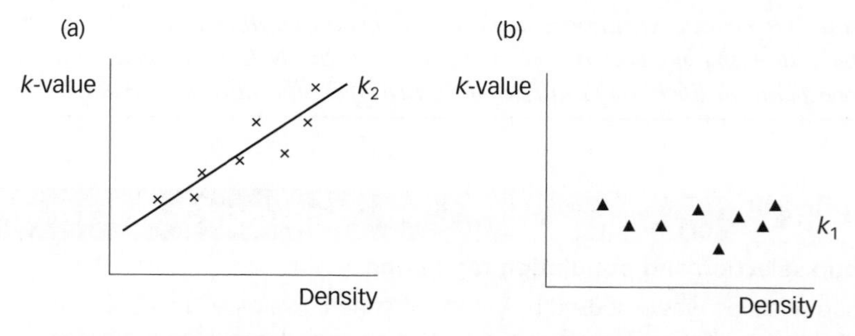

Figure 4.7 Detecting density-dependence in mortality factors from a key factor analysis. The correlation between k_2 and the population density of the second age class (a) suggests this mortality depends strongly on density. k_1 is density-independent (b). Even so, k_1 may be the key factor determining total population size.

Population variability in *r*- and *K*-selected species

- *r*-selected species have little prospect of exhausting their carrying capacity and are density-independent: their mortality rates show no correlation with the number of individuals per unit area.
- A habitat is likely to be first reached by *r*-selected species which will grow rapidly and approach a stable age distribution.
- *K*-selected colonizers grow to their carrying capacity and, if the habitat is long-lasting and predictable, remain there. These populations are density-dependent and approach a stationary age distribution.

Few species are perfectly *r*- or *K*-selected, and some can switch strategies according to conditions in their habitat. Nevertheless this classification is a useful description of alternative life history strategies (and population dynamics) and helps to explain important features of species' autecology.

4.6 OTHER STRATEGIES

The capacity of any individual to maximize its fitness depends not only on its life history strategy but also on other members of its species (competitors and those with whom it cooperates) and with other species (competitors and those with whom it forms mutualisms). It will depend on others it uses as a resource (its prey) and those which use it as a resource (its predators and parasites). We examine models of interspecific relations in Chapter 5.

Strategies develop according to an individual's interactions with other individuals.

- The trade-off strategies in Box 4.1 are principally about finding a mate and the frequency of mating. **Semelparous** species (as well as many **iteroparous** species) have to synchronize their reproductive activity within the population or with other populations.
- Sexual selection between partners can decide the allocation of resources to growth or reproduction. Growing large may be a strategy for securing a limited resource—a mate—or for crowding out competitors for this resource.
- Dominant bull elephant seals weigh close to 4 tonnes, useful when size matters in the competition for the harem on the beach. Large territories can support many mates and the raising of many offspring.
- It may benefit individuals to cooperate to secure a resource. Working together to catch prey, or to watch for predators, will persist if most members of the group achieve a net gain. Female lions within a harem secure most of the food, sharing the risks of bringing down large prey and the costs of protecting the group's young.
- Individuals adopt strategies that give them a selective advantage, whether through cooperation or cheating (Box 4.5), for as long as these are not made disadvantageous by the activity of others.

Box 4.5 **Looking for extra marks?**

Evolutionarily stable strategies

Although the evidence for significant group selection is weak (Section 2.5), in some populations the majority of individuals do adopt the same behaviours and strategies. The obvious question is why they should all do the same thing. Behaviours which confer a significant advantage, especially when individuals cooperate or compete with each other, are likely to be adopted by the majority. Indeed, the advantages of following a particular strategy may depend on others doing the same.

Strategies which persist over generations, resisting challenges from alternatives, are said to be evolutionarily stable. In some cases, two different strategies can persist side by side as long as their frequency in the population allows for co-existence. Following the example in Section 2.5, cheats prosper as long as most individuals do not cheat.

John Maynard Smith used game theory to show how **evolutionarily stable strategies (ESSs)** may establish themselves without the need for group selection. A population dominated by a single strategy is said to have a *pure* ESS. It out-competes all alternatives and individuals with an inferior strategy are soon displaced by those enjoying greater reproductive success. However, two different strategies may persist together if a balance between them can be established in the population.

Consider a population where cooperation between individuals is the norm, but a strategy of cheating also occurs. Cheating may confer a big advantage in encounters with a non-cheat, but it cannot become dominant in the population. We can see this from the possible payoffs from each combination of association:

i. *between two cheats*: both lose out because both cheat. Each incurs costs that are not offset by any benefits;
ii. *between two non-cheats*: each incur the costs of cooperating, but both enjoy a small payoff because the cooperation confers a greater benefit;
iii. *a non-cheat with a cheat*: the large payoff goes to the cheat. The cheat avoids most costs but the non-cheat loses out because any benefits do not cover its costs.

It will be the average payoff from a series of encounters which determine the success of either strategy, and this average will be set by the frequency of encountering a cheat. If *cheat/cheat* encounters are common, the advantage from cheating disappears: the cheat's average payoff declines and with it their reproductive success. Non-cheats benefit from cooperating with other non-cheats (albeit with lower payoffs) but again, this strategy is only viable if there are few encounters with cheats.

Cheating also only works when cheats are few. Frequency-dependent selection will push the population to the maximum proportion of cheats: the level beyond which the number of encounters with non-cheats reduces the average benefit to either strategy. Where this balance lies depends on the relative size of the rewards and penalties from each association: the proportion of cheats in the population will fall as the costs of cheating on a cheat rise. When a balance between the two strategies becomes established, there is said to be *mixed* ESS prevailing in the population.

Of course, non-cheats would benefit if there were no cheats, but their strategy cannot prevent infiltration by cheats.

Any ESS will have the highest benefit-to-cost ratio compared to tested alternatives (tried and tested by natural selection). Any new strategy would need to match or better this advantage and impart a higher fitness to displace the established strategy.

➔ *Section 1.2, Optimality theory; Section 2.5, Types of selection*

This is not group selection, since no advantage is conferred to the group by the strategy, or, at least, this is not implied by the balance of costs and benefits in an ESS. It is, however, a property of the group, since this trait can only establish itself (in our example) in the absence of cheats; or, more generally, given the schedule of payoffs and the frequency of these interactions between individuals.

Check your understanding

Examination-type questions

1. Explain why the largest cause of mortality in a species may not be the main determinant of population size in a key factor analysis.
2. What sort of age distribution would you expect with an *r*-selected species? Why should this distribution indicate an unpredictable or changeable habitat?

You'll find answers to these questions—plus additional exercises and multiple-choice questions—in the Online Resource Centre accompanying this revision guide. Go to http://www.oxfordtextbooks.co.uk/orc/thrive or scan this image:

5 Interactions between species

Key concepts

- An individual may form an association with other individuals or other species to improve its fitness. This association will persist if the benefits outweigh the costs.
- With cooperative associations both species benefit; consumption of one species by another confers a benefit on one and a cost on the other; competition places a cost on both species.
- The Lotka–Volterra model of interspecific competition suggests two species competing for a limited resource only co-exist when both populations are limited principally by intraspecific competition. If the pressure from another species is greater than that from other individuals the competing species will dominate.
- The competitive exclusion principle states that two species with identical demands are unlikely to co-exist in the same habitat. However, the complexity of their shared ecological space may facilitate co-existence through reduced overlap on key gradients.
- Consumption takes a variety of forms according to the nature of the interaction, the impact on the consumed, and the length of the association. This includes herbivory, predation, and parasitism.
- Plants have a variety of structural and chemical means of dissuading herbivores. Higher plants have growth strategies and morphologies that can accommodate the loss of meristematic and photosynthetic tissues to grazers and browsers.

- Animals adapt their behaviour and appearance to defend against predation, including camouflage, protective structures, and weaponry.
- The Lotka–Volterra model of predator–prey populations can cycle through a regular series of oscillations because of the delayed response in each population to the size of the other.
- Host–parasite and host–pathogen models measure the minimum number of susceptible hosts needed to sustain an infection in a population.
- Symbiosis is a range of mutually beneficial associations between two species that illustrate the co-evolution of species interactions.

Assumed knowledge

Basic mathematics.

Note that the models in this chapter work with either absolute numbers or population densities.

A conceptual map for this chapter is available on the companion website; go to http://www.oxfordtextbooks.co.uk/orc/thrive/ or scan this image:

5.1 SPECIES INTERACTIONS

The interactions between species are driven primarily by the need for resources. One or both members of an association attempt to secure a resource—food, space, shelter—with the other species being the resource or the means of acquiring it.

There are three broad categories of association according to the costs and benefits for the interacting species (Box 5.1):

- competition: when a resource is limited, both competitors lose out in the presence of the other;
- consumption: when one species is the exploited resource and may suffer a fatal cost or reduced fitness for the benefit of the consumer;
- cooperation: some associations benefit both partners, or a benefit to one comes at no cost to the other.

These types grade into each other: predation may play a role in some competitive interactions, microparasites share characteristics with pathogens, and so on.

5.2 COMPETITION

Just as individuals within a population compete for limited resources (intraspecific competition), so two species may reduce each other's carrying capacity (interspecific competition).

Competition

Box 5.1 Types of species interaction

Competition

Two or more species use the same limited resource, checking the growth and reproduction of both species.

Exploitative competition is where the use of a resource by one species limits the availability to others.

Interference competition is where one species prevents access to a resource by other species.

Consumption

One species consumes another living species; this improves the growth and reproduction of the consumer but halts or reduces that of the consumed organism.

Predation: carnivory, when one species consumes an animal. This includes plants eating animals. With **cannibalism** individuals eat members of their own species.

Herbivory: when an animal consumes a whole or part of a plant.

Parasitism is where one species lives in extended association with another, deriving resources from its host without killing it. **Parasitoids** are a special group that kill their host at a particular stage in their life cycle.

Disease is a pathogen (viral or microbial) living in a host. The host represents the only ecosystem and means of replication and dispersal for the pathogen.

Cooperation

Two or more species live in association, and both benefit, or one benefits at no cost to the other.

Mutualism: both species benefit from their association. In **symbiosis** the association is so close that each partner suffers if the association is broken.

Commensalism: the benefit derived by one species from the association does not benefit or cost the other species.

Intraspecific competition occurs between members of the same species with almost identical requirements.

- Individuals, differing in size, age, and gender, can have different demands, and different capacities to secure a resource.
- Different life stages may occupy different niches, to reduce intraspecific competition.
- The carrying capacity for a population will be set by one or more limited resources. Intraspecific competition is often the prime driver of density-dependent mortality and fecundity. Survival decreases as resources become scarce.

  ➔ *Section 4.4, Population growth in a limited environment*
- These effects result from both exploitative and interference competition (Box 5.2). Over-crowding can limit the growth and reproductive activity of the weak or tardy in a population.

Box 5.2 Types of competition

Exploitative

Organisms attempt to consume a resource and exclude competitors.

Scramble competition occurs especially between individuals of the same species, where speed of consumption is key.

Pre-emptive competition is the rapid colonization of a resource such that late arrivals, perhaps other species, are prevented from exploiting it.

Interference

Organisms attempt to exclude other individuals or species from access to a resource. This includes the following.

Chemical competition: the production of toxic chemicals or deterrents. Plants may add such deterrents to the rhizosphere and limit germination in their vicinity (**allelopathy**).

Overgrowth competition: the exclusion or limiting of competitors by a rapid growth in size or numbers, most especially in sedentary animals and plants.

Animals may use aggression to defend a resource in **encounter competition**.

Territorial competition operates indirectly by marking and defending a territory, for breeding or feeding.

- Some populations are held below their carrying capacity by their interactions with other species. For them density-dependent intraspecific competition is not significant.
- Individuals may compete for a partner and different types of competition can also be seen when the limited resource is a potential mate.

Revision tip

You will need examples to illustrate all forms of species interaction.

Consider especially detailed examples for the different forms of competition in sexual selection.

Lotka–Volterra model of interspecific competition

Two species competing for a limited resource may not be able to co-exist because of the nature of their competitive interaction (Box 5.2), perhaps because one creates conditions which prevent the other from becoming established.

➔ *Section 6.2, Inhibition in succession*

Alternatively, their co-existence may depend on their relative population growth rates: the speed with which each occupies a limited resource. These interactions are explored with the Lotka–Volterra competition model.

This uses the logistic model of population growth, with rates of increase reducing as population size approaches the capacity of the habitat. Part of this capacity is occupied by a second species so the model incorporates the growth of both species.

➔ *Section 4.4, The logistic equation and its notation*

The model assumes that both species (*a* and *b*) are limited by the same resource and this alone determines their carrying capacities (K_a and K_b). In the absence of a competitor, the rate of population increase of each species is slowed only by *intraspecific* competition ($K - N/K$):

Species *a*

$$\frac{dN_a}{dt} = r_a N_a \frac{(K_a - N_a)}{K_a}$$

Species *b*

$$\frac{dN_b}{dt} = r_b N_b \frac{(K_b - N_b)}{K_b}$$

- In each case, adding another individual reduces the unused carrying capacity for that species by one.
- It will also reduce the capacity for a competitor: we need to add the inhibition due to *interspecific* competition to that of intraspecific competition for each species.
- However, competing species may not need the same amount of the resource: each *a* may, for example, requires twice the resource of each *b*.
- We need to convert the number of one species into the equivalent number of the other, according to their resource requirements. In our example, two *b* individuals will occupy the resource of one *a* individual.
- The rate of exchange is given by **competition coefficients**: the competitive effect of an individual of species *b* on the capacity for population growth by species *a* (α), or, of each *a* on *b* (β).
- In the simplest models, we can derive α as K_a/K_b and β as K_b/K_a. Here, species *a* has a carrying capacity (K_a) of 400, but the same habitat could support 800 species *b* (K_b), so $\alpha = 0.5$ and $\beta = 2.0$.
- In more detailed and realistic models α and β do not have to be reciprocal. (Note that these competition coefficients only apply to the limiting resource being modelled.)
- Multiplying the number of species *b* by α gives the equivalent number of species *a*. This is then expressed as a fraction of the carrying capacity for species *a*:

$$\alpha N_b / K_a$$

And, similarly, for species *b*:

$$\beta N_a / K_b$$

This is the reduced capacity for growth in each species due to interspecific competition. This can be incorporated into their population equations:

Species *a*

$$\frac{dN_a}{dt} = r_a N_a \frac{(K_a - N_a - \alpha N_b)}{K_a}$$

Species *b*

$$\frac{dN_b}{dt} = r_b N_b \frac{(K_b - N_b - \beta N_a)}{K_b}$$

For example, 300 individuals of species *b* would occupy resources that could support 150 *a* individuals ($\alpha N_b = 0.5 \times 300 = 150$). If there was a population of 120 *a*, the occupied space would thus be 270 and the proportion of the carrying capacity for species *a* that remains unoccupied is:

$$(400 - 120 - 150)/400 = 130/400 = 0.325 \ (32.5\%)$$

and 67.5% is occupied.

When the resource becomes completely occupied, no further growth is possible in either population. This happens when a species reaches its carrying capacity:

$$N_a = K_a \quad \text{or} \quad N_b = K_b$$

or its competitor's population reaches a size that matches this capacity:

$$N_b = 800 = K_a/\alpha$$

or

$$N_a = 400 = K_b/\beta$$

or either capacity is reached at some combination of N_a and N_b.

- These are zero-growth points for each species. They can be plotted on graphs of N_a against N_b (Figure 5.1a) and the line connecting these points is the zero-growth isocline.
- Along this line are all the combinations of $N_a + N_b$ where the resource is entirely occupied, with no capacity for growth by that species.
- A second graph can show the equivalent line for the second species. When the competition coefficients are reciprocal, the two lines are identical.
- When they are not reciprocal, plotting the zero-growth isoclines for each species on the same graph indicates the combinations of population sizes where co-existence is possible (Figure 5.1b).

For persistent or stable co-existence, each population must come to a point where growth has ceased and the resource space is exhausted:

$$K_a - N_a - \alpha N_b = 0 \quad (\text{for species } a)$$

and

$$K_b - N_b - \beta N_a = 0 \quad (\text{for species } b)$$

which re-arrange to

$$N_a = K_a - \alpha N_b$$
$$N_b = K_b - \beta N_a$$

- These give the maximum population size of a species, for a given size of its competitor: if N_b is 600, the maximum size of N_a is 100:

$$N_a = 400 - (0.5 \times 600)$$

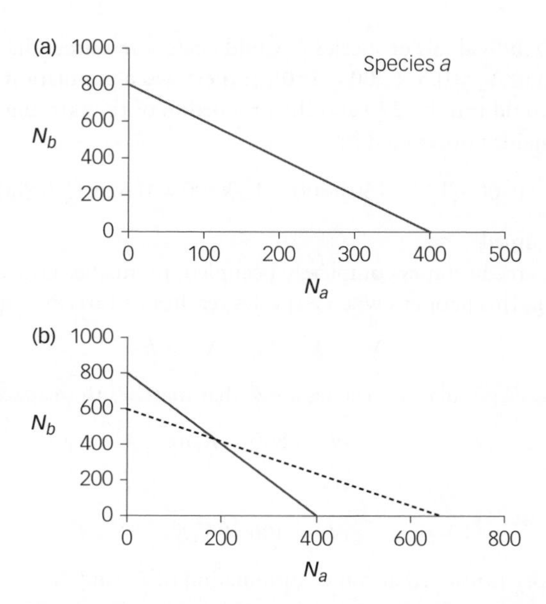

Figure 5.1 (a) The Lotka–Volterra model of competition and the circumstances when one species (*a*, in this case), can no longer grow. The line is a zero-growth isocline, connecting the points where together the two populations completely occupy the limiting resource. This crosses the *x* axis at K_a (species *b* is absent), and the *y* axis at K_b (species *a* is absent).

To the left of this line species *a* can increase (there is unused capacity); to the right the carrying capacity is exceeded (and its numbers must decline). An equivalent line could be drawn for species *b*.

(b) Combining the zero-growth isoclines for both species in the same graph identifies the conditions for stable co-existence. This is where the two lines cross, where both populations can grow no further. A species with a line that sat entirely above the other would eventually exclude its competitors.

Note that here the competition coefficients are not reciprocal.

- These are the population sizes when intraspecific and interspecific competition together prevent population growth. Co-existence is possible since neither species can continue to grow and occupy all of the resource.

The model assumes unchanging environments and fixed values for r_a, r_b, K_a, K_b, α, and β. Despite these highly simplifying assumptions, it does demonstrate the possible outcomes and their conditions:

- species *a* always thrives and species *b* goes extinct;
- species *b* always thrives and species *a* goes extinct;
- the eventual winner depends on the relative size of the starting populations and circumstance;
- the two species co-exist indefinitely.

The outcome depends on the relative growth rates of each population (determined principally by the *K* values and competition coefficients).

- A population will oust the other when it adds individuals fast enough to halt the growth of the other: the species with the greater inhibition of its opponent wins. Or, more precisely, the losing population suffers interspecific competition that is more limiting than its intraspecific competition.
- Thus stable co-existence will not occur if interspecific competition is stronger than intraspecific competition for one species, or when the two species have exactly the same per-capita rate of resource usage.
- The third option occurs when *both* populations are limited more by interspecific than intraspecific competition. This creates an 'unstable equilibrium', a delayed resolution, with the outcome depending on the size of the two starting populations. Given the time needed to resolve the competition, even a small disturbance in the habitat may shift the advantage from one to the other, hence the instability.
- A 'stable equilibrium' will develop when *both* populations are limited by intraspecific rather than interspecific competition. Neither species can achieve a population size that prevents the other sustaining a viable population.

Revision tip

1. Check these equations using the values provided in the example, especially the conditions when no population growth is possible.
2. Worksheet 3 on the companion website provides a model for you to manipulate the population characteristics; go to http://www.oxfordtextbooks.co.uk/orc/thrive/ or scan this image:

Plots of the zero-growth isoclines allow you to explore the conditions where stable co-existence is possible. Both will enable you to become familiar with the population characteristics described in Section 5.2 under the Lotka–Volterra model.

➔ *Section 5.2, Lotka–Volterra model*

Tilman's models of competition over two resources

More realistically, several resource gradients are likely to determine the outcome of competitive battles between species.

- Tilman devised competition models with more than one resource limiting to both species. When the outcome depends on two resources these can be visualized using two-dimensional plots (Figure 5.2). The models incorporate the rates of consumption and supply of each resource and specify the circumstances under which each population would grow.
- These show how the shortage of one resource can limit the exploitation of a second resource by that species. If a different resource limits the growth of each species, co-existence may develop over the long term.

Co-existence between species, even among those with almost identical requirements, is far from rare. In many terrestrial plant or freshwater phytoplankton communities there

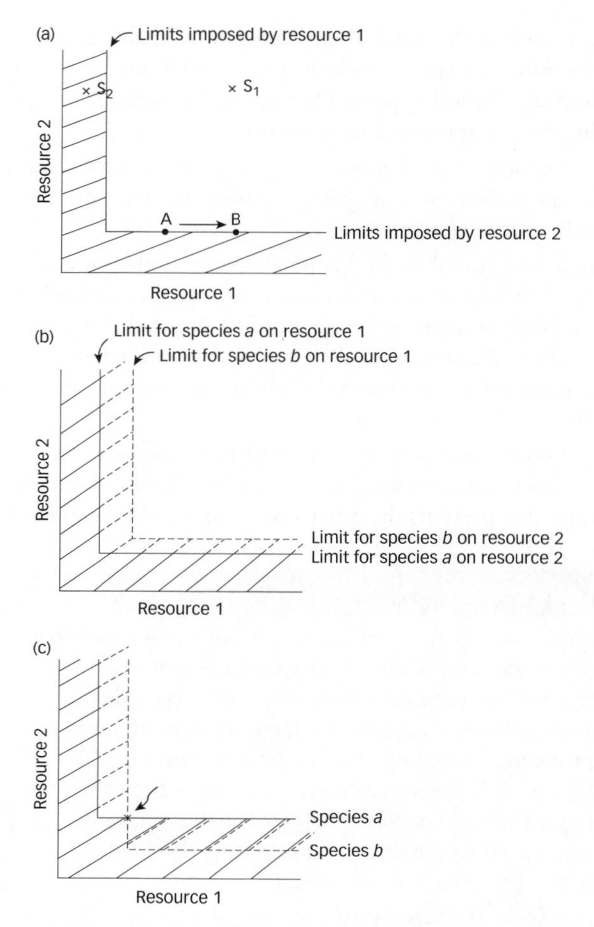

Figure 5.2 Tilman's model of interspecific competition when two resources are limiting. The supply of the two resources and the consumption needed by a species to maintain a viable population is shown on each plot.

(a) For a single species. The zero-growth isocline is shown by the continuous line: horizontally it shows the minimum amount of resource 2 needed and vertically the minimum amount of resource 1. At their junction both resources are limiting: growth is possible above the line, but below it (hatched area) the population declines. If the supply of resource 1 increases more of it could be exploited (from A to B), but the population cannot grow because it is limited by the supply of resource 2.

If the supply of the two resources is at the position indicated by S_1 the population grows. In contrast, at S_2 there is insufficient resource 1 and it must become extinct.

(b) For two species competing for these resources. Species *a* wins because it can maintain a viable population at lower levels of both resource 1 and 2 compared to species *b*. The result is competitive exclusion when species *a* can reduce resource availability below the isocline of species *b*.

(c) Co-existence may be possible when each species can sustain itself on lower levels of one resource compared to its competitor: neither population can grow beyond X (arrow): species *a* is limited by resource 2, species *b* by resource 1.

Given the rate of resource consumption for each species the model predicts the conditions where each is checked by one of the resources.

is little obvious niche differentiation between multiple species. Models of competitive interactions need to explain how such communities can develop and persist, and how competitive battles are defused between species with high niche overlap.

➔ *Section 6.3, Species richness and community structure*

Looking for extra marks?

The simplicity of competition models (and of niche overlap and competitive exclusion) can be usefully contrasted with the detail of published laboratory experiments and field studies.

Competitive exclusion

Based on the large number of laboratory experiments using simple communities, competing species exerting a high competitive pressure are not expected to persist together for long.

➔ *Section 2.2, Interspecific competition*

- The **competitive exclusion principle** states that two species with the same ecological requirements cannot live in the same habitat at the same time. With unchanging conditions, and little immigration or genetic change, one species must eventually disappear.
- The more efficient species—able to maximize its reproductive output with the resources available—is likely to dominate.
- With low niche overlap the intensity of interspecific competition may be small and allow for the co-existence predicted by the Lotka–Volterra model.
- Co-existence becomes less likely where there is considerable overlap over several environmental gradients.
- Sometimes niche overlap may not represent competition because the resource is not limiting. Or potential competitors are separated in ecological time, exploiting the resources at different times. The rate of resource usage and supply is then crucial.
- The question becomes how much overlap will allow co-existence? Or, how much overlap on how many resource gradients? And do other factors reduce their competition?
- Interactions with other, non-competing, species may also facilitate co-existence. Predators, parasites, or pathogens may all check the growth of a population, reducing its competitive effect on another species.
- A population constrained to a marginal habitat, albeit with a low risk of extinction, may adapt to the different ecological space. Such displacement is why competition is regarded as a major driver of speciation.

➔ *Section 2.2, Character displacement*

Box 5.3 Make the connection

Fundamental and realized niche

A species' fundamental niche is the range, across all key environmental gradients, it would occupy in the absence of competitors.

Competition (or some other forms of species' interaction) limits a species to an ecological space some distance away from its optimum. This may apply to just one or to several gradients. If it survives this realized niche, the population will have a reduced fecundity and a higher mortality. The selective pressures may, over generations, prompt character displacement that could lead to speciation.

Being confined to a restricted range but becoming better adapted to it results in resource partitioning. Niches have then become differentiated and species are spaced along the resource gradient.

We may find species co-existing because their competitive battles were fought sometime ago, and have since undergone character displacement.

➔ *Section 1.2, 2.2, Niche and competition*

Revision tip

Note that many of these qualifiers for the competitive exclusion principle refer to mechanisms that reduce the intensity of interspecific competition. This was highlighted as crucial for co-existence in the Lotka–Volterra model.

Few environments remain unchanged for long, and a poor competitor able to sustain itself may eventually flourish when conditions change. This complexity and changeability may help to explain the co-existence seen in many communities.

➔ *Section 6.4, The complexity of established communities*

Looking for extra marks?

Competition in the real world

Look closely at the examples you use to illustrate competition. Evaluate the evidence critically for the nature and strength of the competitive interaction.

Note the difficulty of measuring coefficients of competition outside the laboratory.

5.3 CONSUMPTION

Consumption is the exploitation of one species by another, exerting a cost that can be fatal to the exploited species or reduce its capacity for growth or reproduction. Plants are not defenceless victims of herbivores, and most have growth forms and strategies that accommodate losses to consumers. In the same way, animals have a range of adaptations to reduce their losses to predators and parasites.

Note that our classification of consumption excludes those organisms which consume dead and decomposing tissues. Their fundamental role in nutrient cycling is considered in Section 7.4.

Herbivory

Herbivory, the consumption of living plant material, includes leaf miners, root and stem borers, sap-suckers, and gall-formers (mostly among insects), but there is also a wide range of **frugivorous** (fruit-eating) and **granivorous** (seed-eating) animals.

Additionally:

- **Grazers** feed on grasses and **forbs**, cropping above-ground parts. Consumers of aquatic algae and phytoplankton are also described as grazers.
- **Browsers** feed on young leaves and shoots, especially of trees and shrubs. This includes some insects as well as vertebrates, and again includes large herbivores feeding on aquatic macrophytes.
- **Ruminants** are mammalian browsers and grazers with a four-chambered stomach. This anatomy maintains a microbial culture of cellulolytic bacteria and allows the stomach contents (the rumen) to be chewed for a second time. Other species have symbiotic relations with similar bacteria but without this anatomy (this includes primates and termites).
- **Omnivores** consume both plant and animal material.

Terrestrial plants have a variety of defences against being consumed (Box 5.4).

- The cost of a leaf differs between growth strategies: low in fast-growing species, high in slow-growing species. This, and the speed at which the leaf can be replaced, is reflected in the chemical (and structural) defences.
- Plants have a range of chemical defences, some of which are only produced when herbivory starts, or during vulnerable stages (e.g. seedlings), or in vulnerable tissues (new leaves and buds).
- These are termed **secondary plant metabolites** because they are by-products of essential metabolic pathways.
- Even so, their production incurs a cost and many plants increase their concentration only when they are consumed.
- Fast-growing plants tend to have general poisons which can be produced very quickly, and induced by herbivore activity. These plants have few anti-feedant chemicals but replace consumed leaves quickly.
- Slow-growing plants produce high levels of anti-feedants but low levels of general toxins. Anti-feedants make the tissues unpalatable or undigestible. They are produced slowly throughout the plant's growth, but higher levels are induced by herbivory.
- The resource availability hypothesis suggests defences against herbivory reflect resource supply. Fast-growing generalists (*r*-selected plants) have inducible defences, and only bear these costs when and where necessary (such as defending the meristematic tissues).
- Slow-growing species may also concentrate their defences where most needed—in new leaves and buds—but have anti-feedants present in all plant parts.

Box 5.4 Plant defence mechanisms

Structural

External features: hairs, thorns, spines, hard external surfaces (including silicate deposits) to dissuade feeding, exudates as traps for insects or other small herbivores.

Adjusting growth pattern (compensation) to restore the balance between root volume and leaf area after consumption.

Chemical

Stings and irritants to dissuade touching or consuming the plant.

Toxins and anti-feedants: not only distasteful chemicals but also those making the tissue hard to digest or which kill the consumer.

Disruptors: chemicals that interfere with physiological processes such as reproduction or moulting in the consumer.

Antimicrobial agents.

Biological

Mutualistic associations where an animal defends a plant against herbivores in return for shelter or access to nutrients.

- The production of thorns and spines by some trees and bushes depends on the browsing intensity from large mammals; in its absence, these structures become less prevalent. Some rapidly growing species also adopt such devices.
- The growth form is also part of the response to herbivory; by storing resources underground, by keeping their leaves or meristematic tissue close to the ground, plants protect their capacity for future growth. Generally, these traits are not induced by grazing activity, but result from a history of adapting to grazing pressure.
- **Compensation** is the replacement of cropped tissues and changes in the growth form due to herbivory, such as growing lateral shoots. Along with the production of secondary plant metabolites these are all inducible phenotypic adaptations.

  ➔ *Section 2.1, Adaptation and acclimation*

Many herbivores, even ruminants, are opportunistic carnivores, especially when short of key nutrients. Similarly, some carnivores supplement their diet sporadically with plant material. There are also consumers which change their diet and their digestive physiology between stages in their life cycle.

Looking for extra marks?

Many secondary plant metabolites have immense pharmaceutical and economic value. You may need to know the metabolic pathways that produce the major groups.

Types of predator

Predators generally consume many prey, many parasites feed on one host, and one parasitoid feeds on one host. Predators kill their prey, parasitoids kill their host (eventually), whereas most parasites and pathogens exploit some part of their host without killing them. Parasitoids represent a transition between predator and parasite, and parasites grade into pathogens.

Models of competition measure the indirect effects of two species on each other. Predator–prey interactions are direct: the resource is the prey and prey numbers decline when it is consumed by a predator.

Animal defence mechanisms against predation are summarized in Box 5.4.

Predator–prey models

The Lotka–Volterra model for predator–prey relations uses equations for the consumer and the consumed populations together. The predator population will grow if they consume enough prey; prey numbers must eventually decline if predator numbers rise.

The model makes several simplifying assumptions.

- Neither prey nor predator experience density-dependent population growth. Both populations change continuously.
- There are no effects of age or sex. Each predator is equivalent, as is each prey.
- Predators consume only one prey species and their population growth is directly proportional to the number of prey they consume.
- Starvation, a shortage of prey, is the only cause of predator mortality.

In the absence of predators, prey numbers (N) grow exponentially, as in an unlimited environment:

➔ *Section 4.4, Models of population growth*

$$\frac{dN}{dt} = rN$$

In the absence of prey, predator numbers (P) decline exponentially: there can be no births, only deaths:

$$\frac{dP}{dt} = -mP$$

(remember, $r = b - m$; with no births $r = -m$: the per-capita rate of population 'growth' equals the mortality rate).

- The number of opportunities for a predator to consume a prey will increase with the abundance of predators and of prey. Assuming that such encounters occur randomly, this can be measured simply as NP.
- However, not every encounter will result in the death of a prey: some will escape. The proportion of encounters leading to a death is the attack rate α. Here α is fixed.

Box 5.5 Animal defence mechanisms

Appearance

Crypsis is merging into the background using colour, pattern, and behaviour (camouflage).

Aposematic colouration uses colours and patterns that predators associate with stings, poisons, or being distasteful.

Polymorphism is where individuals adopt a different appearance from most of their species to go unrecognized by predators.

Mimicry

Batesian mimicry: adoption of the appearance of an unpalatable or dangerous organism by a species that doesn't have these traits.

Mullerian mimicry: a collection of dangerous or unpalatable species evolving similar appearance independently, which together reinforces the effect of their aposematic colouration.

Behaviour

Aggression: intimidating displays and fights engaged with predators, either singularly or collectively by mobbing.

Flocking, herding, and shoaling: individuals reducing their risk of predation by being a part of a large group: through chance, intimidation, or confusing the predator.

Signalling: calls or displays that alert a group about a potential predator.

Distraction: directing the predator's attention away from eggs or young.

Biology

Masting: overproduction of offspring to overwhelm the functional response of a predator.

Weaponry and armour: protecting vulnerable areas by reinforced skin or plates or spines or bearing intimidating weapons.

Chemical defences: smells, stings, and sprays that deter; poisons which reinforce aposematic colouration.

Shedding tissues: distracting the predator by leaving tissues behind on which it can feed, while the owner escapes.

- αNP thus gives us the number of prey taken by the predator. For example, with $N = 200$ and $P = 20$ there will be 4000 encounters. Given an attack rate of 0.01, then:

$$\alpha NP = 0.01 \times 200 \times 20 = 40 \text{ prey are killed}$$

This loss has to be subtracted from the prey population. This is the mortality rate due to predation:

$$\frac{dN}{dt} = rN - \alpha NP$$

Predators that catch prey can give birth, and their birth rate is determined by the number of prey they catch.

- The rate at which prey can be 'converted' into predator births is given by β. This is how many prey have to be consumed to support the birth of one predator.

For example, if five prey are needed for one birth then $\beta = 0.2$ and:

$$\beta(\alpha NP) = 0.2 \times 40 = 8 \text{ predators are born.}$$

The predator now has a birth rate—$\beta(\alpha NP)$—so its population can grow:

$$\frac{dP}{dt} = \beta(\alpha NP) - mP$$

(that is, $r = \beta(\alpha NP) - m$)

- Again, the assumption is that β remains fixed.

A time series of these simple models produces a series of cycles (Figure 5.3a), with the fortunes of the two species locked together. The populations are said to show a 'periodic solution'.

- When the birth rate of the prey cannot match its losses to the predator, N must decline.
- Falling prey numbers support fewer predator births, so P must eventually fall.
- Delays in the response of each population to changes in the other population maintain these oscillations.

The Lotka–Volterra model again shows when prey or predator checks the opponent's population growth. The prey cannot grow when additions are matched by losses to the predator:

$$rN = \alpha NP$$

Dividing throughout by N:

$$r = \alpha P \quad \text{and} \quad P = r/\alpha$$

That is, the size of predator population needed to halt prey growth is given by the intrinsic growth rate of the prey divided by the attack rate. If we set r to 1.2, then in our example ($\alpha = 0.01$), zero prey growth happens when the predator population reaches 120.

An increase in the attack rate will halt prey population growth at a lower predator population size.

Similarly, the predator will cease to grow when:

$$\beta(\alpha NP) = mP$$

and

$$\beta(\alpha N) = m \quad \text{and} \quad N = \frac{m}{\beta\alpha}$$

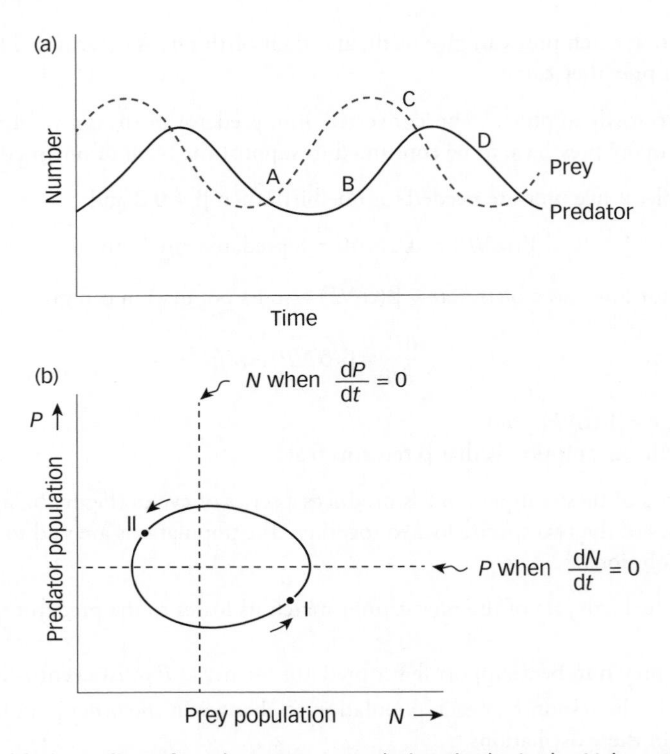

Figure 5.3 (a) The cycling of predator–prey populations in the Lotka–Volterra model. At A prey increase because predator numbers are small. Rising prey numbers allow the predator to increase at B. The predator begins to check prey population growth at C, and as prey number in turn decline, predator numbers follow. The cycle begins again. (b) The cycling of predator and prey populations that never resolves. Each position along the loop marks the size of *both* populations. The cycle progresses in the direction of the arrows.
The zero-growth lines are indicated by the dashed lines: when predator numbers are below the horizontal line, the prey population can grow; when the prey numbers are to the right of the vertical, the predator population can grow.
At position I predators increase because prey numbers are high (and growing). They increase until the trajectory crosses the vertical (when prey numbers fall below the equilibrium position). At position II prey numbers fall because predator numbers are above their equilibrium. The prey increase again when the trajectory crosses the horizontal and predator numbers are below the equilibrium. This graph could be redrawn for different ratios of the two populations, creating a series of such loops.

This is how small the prey population must become to halt growth in the predator population. In our example (setting the mortality rate for the predator to 0.4) this is:

$$N = \frac{0.4}{0.2 \times 0.01} = 200$$

With a more efficient conversion of prey into offspring or a higher attack rate, the predator can maintain its population with a smaller prey population.

Again the interaction between the two populations can be explored through zero-growth isoclines for each population, when the size (or density) of the predator is plotted against that of the prey (Figure 5.3b).

Explore these cycles with the model in Worksheet 4 on the companion website; go to http://www.oxfordtextbooks.co.uk/orc/thrive/ or scan this image:

These cycles can be found in nature but it is difficult to create them in the laboratory. Simple bench-scale ecosystems invariably lead to the predator eating all the prey and thereafter starving to death.

With the exception of parasitoids, few predators respond to rising prey numbers by simply increasing their birth rate. A larger prey population allows each predator to consume more prey per unit time (termed the **functional response** of the predator) and later produce more offspring (the **numerical response**).

- The functional response is given by αN in the Lotka–Volterra model, the number of successful attacks.
- The model makes no allowance for predators becoming satiated. For most predators satiation would mean the functional response is not linear.
- The numerical response is given by $\beta(\alpha NP)$, the increased birth rate. Again, this will not be linear for most predators because there are other limits on their rate of reproduction.

Time lags in the functional response (such as learning the prey image) and the numerical response can dampen predator–prey cycles.

- How quickly an abundance of prey is translated into new reproductive predators depends on the predator's generation time, and also the number of existing young, the survival of which improves with increased food.
- Natural ecosystems are not uniform but have patches where prey may escape predation. Searching and handling times vary according to where the predator is feeding.
- To reduce its search costs, a predator will stay where prey are abundant or chase those easily caught.
- With alternative prey species available, the predator's choice should reflect the cost of finding and handling each species and also the benefit represented by each, either in terms of energy or key nutrients (Box 5.6).

Adding refuges for the prey allows oscillations to develop in laboratory cultures but they rarely last more than a few generations, before one or other population goes extinct. Other details confound any simple relationship between predator and prey numbers.

- Each individual is not equivalent. Some predators are better killers than others and some prey are more likely to get eaten. Predators often take the weak and the infirm, including the youngest and oldest which may not be reproductively active. Their loss does not lower the capacity for prey population growth.

Box 5.6 Make the connection

Optimal foraging theory

The costs associated with finding and handling prey need to be weighed against the benefits gained with each catch. Optimal foraging theory applies cost-benefit analysis to food acquisition, to determine when the consumer should switch its attention and effort.

Consider a predator hunting two species of prey, *a* and *b*. The benefit from each prey consumed is designated E_a and E_b. Against this are the costs of capturing and handling an individual: h_a and h_b respectively. The relative reward for each prey is thus E_a/h_a and E_b/h_b.

We need to include the time taken to locate each prey: the rates of encounter per unit time are λ_a and λ_b. Over a search period of T_s the predator would encounter $T_s\lambda_a$ species *a* and $T_s\lambda_b$ species *b*.

The handling time for each prey multiplied by the number caught gives the time spent consuming the total catch: $T_s\lambda_a h_a$ and $T_s\lambda_b h_b$.

The total reward (E) to the predator is the product of the number of encounters and the benefit derived from each prey species, added together:

$$E = T_s\lambda_a E_a + T_s\lambda_b E_b$$

or

$$E = T_s(\lambda_a E_a + \lambda_b E_b)$$

The cost is the total time taken to acquire E; that is, the sum of the search and the handling times for each species:

$$T = T_s + T_s\ (\lambda_a h_a + \lambda_b h_b)$$

So

$$\frac{E}{T} = \frac{\mathrm{T_s}\ (\lambda_a E_a + \lambda_b E_b)}{T_s + T_s(\lambda_a h_a + \lambda_b h_b)} \quad \text{or} \quad \frac{\lambda_a E_a + \lambda_b E_b}{1 + \lambda_a h_a + \lambda_b h_b}$$

This is the reward, per unit time, from consuming both prey species as they are encountered.

For species *a* alone, this is

$$\frac{E_a}{T_a} = \frac{\lambda_a\ E_a}{1 + \lambda_a h_a}$$

When species *a* provides a greater return than *b*:

$$\frac{\lambda_a E_a}{1 + \lambda_a h_a} > \frac{\lambda_a\ E_a + \lambda_b E_b}{1 + \lambda_a h_a + \lambda_b h_b}$$

a predator should chase prey *a* alone, rather than both.

This is a very simplistic (and unrealistic) model of predator behaviour. It assumes that both prey species are encountered with the same frequency and have similar handling times.

Most predators do not switch prey species as readily as the model would predict. However, it does help to identify the important elements determining the rate of reward, especially the costs of searching and handling, and when predators should give up the chase. These costs can be vied between each component: if search times are high, handling times have to be low (or vice versa), otherwise the costs of each kill will be high.

- Density-dependent effects (intraspecific competition) are likely to have an impact on the growth of both prey and predator populations. Predation pressure which is not excessive can reduce intraspecific competition and increase the capacity for growth in the prey population.
- If a predator exploits a single prey (as with many parasitoids) some prey will have to escape if both species are to persist.
- Many predators feed on several prey species and can maintain themselves when one species is in short supply. Predators tend to concentrate on the most abundant species to reduce their search costs.
- Many also compete with other predators for some prey species. This weakens the link between the growth or decline of one predator with prey species abundance.
- Finally, an oscillation in prey numbers may be generated by other factors in their habitat: real-world predator–prey interactions often show us that some extrinsic factor is needed to explain the persistence of such cycles.

Looking for extra marks?

Use this list of complicating factors to critically evaluate your examples, both in laboratory experiments and field trails that attempt to follow predator–prey population dynamics.

Parasitoids, parasites, and pathogens

This group comprises species that use one or more organisms as a habitat and its only source of nutrients.

- **Parasitoids** (principally insects of the orders Hymenoptera and Diptera) grow as a larva in a host. Usually there is one parasitoid in each host and it kills the insect host when it is ready to emerge. The adult parasitoid is free-living and dispersive.
- Many **parasites** infect several hosts: a **definitive host**, where it reproduces, and often one or more intermediate hosts used to disperse. Each host may contain many individuals of a single parasite species, and in most cases will survive the infection.
- Parasitic plants are either hemiparasitic—deriving some of their nutrients from their host plant but able to survive without it—or holoparasitic—obligate plant parasites that may have no capacity to photosynthesize themselves.

Consumption

- A **pathogen** is any organism which causes a disease in its host. Parasites and pathogens are not always clearly distinguished: *Plasmodium* is the parasitic protozoan that causes malaria, whereas many humans carry tapeworms such as *Taenia* with no outward sign of their infection.

Revision tip

You are likely to study examples of one or all of these associations. Ensure that you can relate the details of their interaction to the ecology of both species. Consider especially how the life cycle of the consumer is related to the autecology of the host.

Models of infectious disease

The simplest model of the transmission of pathogens applies to viruses, bacteria, and protozoan parasites with a short life cycle, reproducing and spreading quickly. It defines the conditions under which an epidemic can develop and sets out the principles that underpin **epidemiology**.

The *SI* model divides the host population (N) into two groups: infected (I) and those susceptible (S) but not yet infected. Individuals that recover re-join the S group. We assume that:

- all hosts are equal and able to transmit the disease as soon as they become infected. No individuals are infected at birth;
- there is no density-dependent feedback among the hosts other than parasite-/pathogen-induced mortality;
- infections occur in direct proportion to the number of encounters between infected and susceptible individuals. As a result, transmission is density-dependent;
- all individuals give birth at the same rate (b).

Again, we run an equation for each of the two sub-populations.

The *susceptible group*, S, will increase with births, or by individuals recovering from infection.

- Both S and I individuals can give birth, so recruitment due to births is $b(S + I)$.
- The rate of recovery per infected individual is v, so vI is the number of recovered individuals.
- Thus S will increase by the sum of the births and recovered individuals: $b(S + I) + vI$.

S will decrease due to deaths, and by individuals becoming infected:

$$m = \text{mortality rate per individual (due to causes other than the disease)}$$
$$\beta = \text{transmission rate (infection rate per encounter)}.$$

- As with the predator–prey model, there will be SI encounters between susceptible and infected individuals, of which a proportion, β, will lead to a new infection: βSI gives the number of S becoming I.
- S consequently falls with non-infection mortalities and losses to I:

$$mS + \beta SI$$

Collecting these together gives the rate of change in the size of S:

$$\frac{dS}{dt} = b(S + I) + vI - mS - \beta SI$$

The *infected group*, I, increases with the new additions: βSI.

- I declines when they die:
  α = mortality rate per individual (due to the disease),
  and m (due to some other cause)
  or recover (v).
  So $(\alpha + m + v)I$ gives the total loss from the I group.
- The rate of change in I is thus:

$$\frac{dI}{dt} = \beta SI - (\alpha + m + v)I$$

For the change in the total population size, N (that is, $S + I$), the two equations can be combined:

$$\frac{dN}{dt} = (b - m)N - \alpha I$$

(N multiplied by the difference between births and natural deaths (= r) gives the change in N, from which is subtracted αI, the additional mortalities due to the infection.)

Many key parameters depend on the life cycle of the pathogen and the time taken for an individual to become infective. These can be estimated using the model.

- The **net reproductive rate** (R_I) *of the disease* is the number of individuals infected by each infected individual. Dividing the equation for the change in I throughout by I we derive the increase in I per infected individual:

$$\frac{1}{I} \times \frac{dI}{dt} = \beta S - (\alpha + m + v)$$

- The disease can only spread if gains (βS) are greater than losses ($\alpha + m + v$). The net reproductive rate for the disease is thus:

$$R_I = \frac{\beta S}{(\alpha + m + v)}$$

That is, the rate of increase of infected individuals, divided by the rate of decrease of infected individuals. The disease will only persist if R_I is equal to or greater than 1.0: each infected individual must pass on the disease to at least one other individual before dying or recovering.

The size of S is critical if the disease is to persist in the population. The **threshold density** (S_T) is the minimum value of S when $R_I = 1$.

- We can find the value of S when $R_I = 1$ by rearranging the above equation:

$$\frac{1}{R_I} = \frac{(\alpha + m + v)}{\beta S} = \frac{1}{1} = \frac{(\alpha + m + v)}{\beta S}$$

multiplying throughout by S

$$S = \frac{(\alpha + m + v)}{\beta}$$

This is the value of the threshold density, S_T.

- Below this density the disease will not persist ($S < S_T$).
- The disease can only spread when $S > S_T$.

Since $1/S_T = \beta/(\alpha + m + v)$ we can see that $R_I = S/S_T$.

So, given the rate of spread (the net reproductive rate) and the number of susceptibles, we can work out their threshold number (density):

$$S_T = S/R_I$$

This is the number of infected individuals needed for the disease to spread.

These models thus help to predict when a disease becomes epidemic.

A single transmission rate for all uninfected individuals, and the assumption that all are equally at risk of contracting the infection, makes the model unrealistic. More detailed models allow for such variations, and for the gain and loss of immunity.

5.4 COOPERATION

Mutualism is a long-term association in which both partners benefit. A close association implies a period of evolution together, where each species has been a significant factor in the ecological space of its partner.

Mutualisms extend beyond the exchange of nutrients (trophic mutualisms). Various species of ant defend the supply of nutrients they harvest from plants or insects (most especially aphids), but, in doing so, defend the associate from herbivores or predators (defence mutualism). The over-production of pollen or seed allows for losses to consumption by the dispersing agent, such as bees consuming pollen and squirrels consuming (and burying) acorns (dispersal mutualism): the incurred costs of over-production are set against the benefits of exchanging genes with distant individuals.

- **Symbiosis** includes obligate associations in which two species persist together, invariably exchanging resources, where either partner is rarely found alone; for example, mycorrhizae.
- Some symbioses may only develop under certain conditions; for example, *Rhizobium* and nitrogen fixation in legumes.

We can find examples of both specialist and generalist symbiotic partners in the **rhizosphere** of plants.

➔ *Section 7.4, Nutrient transfer in terrestrial ecosystems*

Box 5.7 **Looking for extra marks?**

Compare the details of symbiotic associations

You will need to know detailed examples of species' interactions and be able to compare similar associations.

Consider the associations between nitrogen-fixing bacteria (*Rhizobium*) and the roots of legumes and between mycorrhizal fungi and the roots of a wide range of plants. In both the host provides a protective habitat and energy-rich carbohydrates for its symbiont: the plant benefits from nitrogen fixed by the bacteria, and phosphates (and other nutrients) scavenged by the fungi.

Most species of *Rhizobium* are associated with a single group of plants which provide oxygen-free nodules on their roots, the conditions under which *Rhizobium* will fix nitrogen if supplied with energy. Outside of this group, *Rhizobium* will form a looser association with a host, without nodules. Mycorrhizal associations are far less specific, and a tree may have several different fungal symbionts within its root system. Many endomycorrhizae require the development of special tissues to support the symbiont; for example, vesicular-arbuscular mycorrhizae grow in vesicles where a close union forms between fungal hyphae and root cells.

Some fungi (e.g. *Frankia*, Actinomycetes) will also fix nitrogen in root nodules in some trees (e.g. *Alnus*). So nitrogen-fixing symbiosis has evolved independently at least twice and mycorrhizal associations have evolved several times.

Despite their lack of specificity, endomycorrhizal fungi are obligate symbionts, needing a host plant to grow and reproduce. Many trees are less productive or fail to establish on soils without fungal partners. However, a host will have fewer mycorrhizae if nutrients are easily available. Similarly, *Rhizobium* and its host do not form a close association if nitrogen is abundant in the soil.

In both symbioses one or both partners will thus avoid the costs of the association if it is not required.

Compare two or more associations for the evidence of co-evolution and suggest reasons why some are closer than others; use the detail of your examples to support your arguments.

- **Commensalism** is where one partner benefits at no cost to the other. Pollination is an example if the go-between does not consume the pollen.
- Commensalisms involve much looser associations, which are often periodic and not always necessary for the survival of the beneficiary.

As an ecological community develops, the activity of some species facilitates the establishment of others. For example, the decomposition of organic matter in the soil involves a microbial succession where cellulose-degrading fungi release sugars that other fungi and bacteria can utilize.

➔ *Section 6.2, Facilitation in succession*

5.5 CO-EVOLUTION

Any form of close association requires one species to become adapted to the presence of the other. Co-evolution would imply that significant changes in the genotype of one species could be matched, over time, with shifts in the genotype of the associated species.

Co-evolution between species has its counterpart, between individuals, in sexual selection. Here a potential mate can produce extravagant phenotypes and undergo rapid adaptive change.

➔ *Section 2.5, Sexual selection*

- The biotic environment and its selective pressures are likely to be changing rapidly. Tracking a partner, and adapting to it, may limit or compromise associations with other species and become part of the trade-off made by a species.

  ➔ *Section 1.2, Trade-offs*
- Co-evolution may extend over the longer evolutionary history of both species. Thus the adaptive radiation of some parasites has closely followed the speciation of their hosts so that their phylogenetic history matches the pattern seen in the hosts.
- Over the shorter term, the development of resistance to a virus by the host can elicit a rapid change in the genotype of the virus.
- In predator–prey interactions such close co-evolution has been termed the 'Red Queen hypothesis': each species needs to keep changing to maintain its reproductive success, and match change in the other.

Such effects could explain the prevalence of sexual reproduction, and may, perhaps, be its principle selective advantage.

- Parasites or pathogens with short generation times can change their genotype faster than their host. However, they will be adapted to the most common genotypes in the host population, those they are most likely to encounter.
- Hosts that cannot be identified are less likely to become infected. Sexual reproduction produces novel genotypes, variation that goes unrecognized by infective agents.

- The same may be true of predators and the time they take to recognize the image of different prey.

➔ *Section 2.4, Heterozygote advantage; Section 2.5, Frequency-dependent selection*

Looking for extra marks?

Some of these associations represent useful examples of disruptive and frequency-dependent selection (see Section 2.5). A detailed case study could usefully show how a rarer phenotype might enjoy a selective advantage, as a result of co-evolution.

➔ *Section 2.5, Sexual selection*

Predation and competition can accelerate change in the gene pool: those that get eaten or don't get to eat do not get to reproduce. These interactions thus have implications for the structure of the larger community.

➔ *Section 8.2, Species diversity in the tropics*

Check your understanding

Examination-type questions

1. What effect would satiation have on the functional response of a predator? What other effects could alter the functional response of a predator to an increase in prey numbers? Illustrate your answer with examples.
2. Why might the phylogeny of a host be mirrored in the phylogeny of some of its parasites? Suggest some possible measures of the closeness of their association from the phylogeny of the host and the parasites.

You'll find answers to these questions—plus additional exercises and multiple-choice questions—in the Online Resource Centre accompanying this revision guide. Go to http://www.oxfordtextbooks.co.uk/orc/thrive or scan this image:

6 The ecology of the community

Key concepts

- Are communities loose assemblages of species, adapted to the predominant abiotic conditions, or the result of extended co-evolution between its members, creating a highly integrated community?
- By observing community change in space and time we can assess whether particular combinations of species result in stable and persistent communities.
- Pioneer species in primary successions show *r*-selected characteristics; late-successional or climax communities are dominated by *K*-selected species.
- Early colonizers of a habitat may promote conditions allowing other species to establish (facilitation) or, alternatively, exclude later arrivals (inhibition). Tolerance to low nutrients or other stressful conditions are characteristics of species in established communities.
- A climax represents a stable assemblage, with low species turnover. However, there is unlikely to be one species assemblage found in all locations; rather, different species occupying equivalent niches.
- The intermediate disturbance hypothesis suggests the highest species richness, *S*, occurs when there is a turnover of species induced by occasional

disturbances which limit competitive inhibition. Many communities are a mosaic of patches at different stages of recovery from a disturbance.

- *S* increases with the area sampled. Models of colonization and extinction in isolated habitats suggest that *S* will eventually reach a dynamic equilibrium where these two processes are in balance.
- Rules of assembly have been suggested to constrain species combinations as a result of their interactions. Such rules are more likely to apply to functional roles rather than species.
- Species are grouped into guilds, groups carrying out similar ecological roles. Functional redundancy is the capacity of an ecosystem to maintain an ecological process when some species are lost.

Assumed knowledge

Basic mathematics and statistics, including tests of association.

A conceptual map for this chapter is available on the companion website; go to http://www.oxfordtextbooks.co.uk/orc/thrive/ or scan this image:

6.1 QUESTIONS OF INTEGRATION

The close association of predator with prey, of host with symbiont, prompts questions of whether these interactions determine the composition of the larger ecological **community**. Beyond their shared adaptations to the abiotic environment, how closely are species tied to each other, to their biotic environment?

Combinations of species repeated in different locations, or unchanged over time, would indicate that some assemblages are more likely to persist than others. It may be that the loss of one species might cause the loss of others or impair some ecological function.

- Even if communities do not have exactly the same species, are only certain combinations of species found together? Are there constraints on which species can remain together?
- Or should communities be compared on the basis of their niches rather than their collections of species?

We can classify communities by their major abiotic differences—for example, aquatic or terrestrial—and subdivide these by further criteria, such as freshwater or marine, **lotic** and **lentic** freshwater systems, etc. Each step more closely defines the key selective pressures of the abiotic environment. We also need to define the biotic pressures and a species' role in its community.

Revision tip

Communities are complex systems and you need to be clear about the aims of studies you use as examples, and the extent to which the research answers some of these questions. The section entitled Two views of community integration compares two classical models of community structure, which you may be asked to review using data from a particular ecosystem.

Note also that this topic is an important interface between the autecology of a species and its community.

Two views of community integration

Early work on ecological communities centred on terrestrial plant communities, not least because of the ease of counting species and following their development using fixed plots. Two alternative views of community integration emerged.

- The organismic or monoclimax model of Clements saw the community as a highly integrated assemblage of species, a product of their co-evolution.
- The prime abiotic factor, the regional climate, determined the dominant plant species and these created a distinct biotic environment.
- Clements suggested its close associations caused the community to function rather like an organism, flexing with minor abiotic changes, but returning to a stable configuration. Certain combinations of species were always likely to be found together and a single **climax community** would be found in a particular climate.
- Gleason and Ramensky argued that communities were much looser assemblages: species were found together because they were adapted to the same abiotic conditions. The assemblage was due, in large part, to chance and the history of the site.
- For Gleason, an assemblage reflected the adaptations of its constituent species and their chances of migrating, colonizing, or establishing a population. Communities would not form distinct units, but have a range of possible configurations.
- This was termed the continuum (or individualistic) model because communities should grade into each other, changing as abiotic conditions changed, with no single regional climax community.

There are various indicators by which the two models can be judged.

The organismic model predicts that:

- discrete communities, characterized by one or more dominant species, will be repeated where there are similar abiotic conditions;
- the same or similar collection of species will be found in different locations;
- the arrival of many species will depend on the presence or absence of other species;
- the fortunes of one species will be closely tied to its neighbours;

- communities will have high elasticity, returning to the same species composition following a disturbance.

The continuum model predicts:

- the community will be variable in time and between locations;
- communities grade into one another along a major gradient: distinct **ecotones** should be rare;
- different species occupy the same niche in different locations;
- consequently few species will disappear if a particular species is lost;
- communities will have a low elasticity: they will be unlikely to reform with the same composition after a disturbance.

➔ *Box 7.3, Concepts of stability in ecology*

Revision tip

Support your answers on community integration with detailed examples and reach a judgement on the two models.

The points listed above provide a checklist to review your examples.

The two models are now regarded as extreme positions. Variations have been proposed, informed by data from a greater range of communities, and identifying various complicating factors:

- some species have greater significance for the development or the structure of an ecosystem;
- associations that were important as the community developed may no longer structure it; conversely, past competitive battles may be more significant than current interactions;
- some associations are closer than others, and have different 'strengths' across the community;
- species and their interactions have evolved to accommodate particular frequencies of disturbance. Beyond the highly predictable seasonal or other short-term changes, there are longer-term changes that cause a turnover of species.

➔ *Box 5.7, Symbiotic nitrogen associations in terrestrial plant communities*

The sequence of arrivals and extinction of species as the community develops—a succession—allows us to measure the predictability of a species assemblage (Box 6.1).

Box 6.1 Key technique

Studying successions and community integration

Some successions allow us to study how a community of species is assembled. A walk down the valley of a retreating glacier can represent several hundred years of community development, from the scoured rubble at the foot of the ice tongue,

continued

exposed last year, to the 300-year-old woodland growing on the moraine several kilometres away. Although sampling in space does not always represent samples in time, we can observe the colonization by pioneer species and their capacity to facilitate and inhibit the arrival of later species down the valley.

We can, occasionally, follow some patches in real ecological time. The ash and lava that created the island of Anak Krakatoa in 1930 has been growing ever since and its succession followed by ecologists. Replicated plots on the mudslides and lava flows of Mount St Helens allow us to measure variation and chance in successional processes.

On such sites we know the history and can follow the arrivals and losses. We can also manipulate plots (planting or removing early or late successional species, adding nutrients, inoculating the soil with microbes, and so on) to study the role of different species on the sequence. Equally we can disturb or disrupt a site and measure the response of the community.

Alongside changes in the species list and the development of the food web, percentage cover, or relative abundance of different groups, ecologists collect data on the development of ecological processes; for example, the development of the soil, its stratification and depth, organic content, nutrient capital, rates of nutrient loss, and so on. Again, plots can be manipulated to interrupt or prevent the development of associations and interactions, for example by examining the effect of fire or grazing pressure on the development of a community.

Modifying replicated plots gives important insights into the integration of a community. They also help us to decide priorities for conservation or the reclamation of degraded communities.

Revision tip

Understand the terminology of climax communities

Order your revision of succession by the models proposed:

monoclimax: Clement's organismic theory of limited combinations of species determined by climate and forming a highly integrated whole;

polyclimax: Tansley's observations that a range of possible climax communities were found within a climate zone, with different plant species dominating;

climax-pattern hypothesis: Whittaker's view that multiple climax communities exist according to local conditions and chance, and these grade into each other. This blends the *continuum* model of Gleason and the polyclimax model, recognizing that many communities consist of a *mosaic* of local patches, cycling through several states, according to their nutrient balance and the condition of one or two keystone or dominant plant species.

You should have at least one detailed example of the *patch dynamics* in a mosaic community you have studied.

6.2 SUCCESSION

Succession is the progressive development of a community through time, from an unoccupied substrate to an established community. Successions can be classified by two criteria, as follows.

By the resources available at the outset:

- **primary succession**: no organic matter is present and major nutrients are present only in mineral form. Consequently, there is little or no multicellular life and, in some cases, no microbial community either;
- **secondary succession**: organic matter is present. A disturbance has destroyed much of the previous community although a microbial community is likely to be present and possibly a seed bank or other propagules.

By the processes that sequence the colonizers:

- an **autogenic succession** is governed by species interactions;
- an **allogenic succession** is governed by external factors, including rates of disturbance or change in some abiotic factor.

- Plant communities on a lava flow or a glacial moraine are classic examples of a primary succession, likely to begin as allogenic successions; a mud flow or a cowpat colonized by invertebrates are secondary successions which begin and end as autogenic successions.
- A secondary succession must inevitably involve autogenic processes, given the presence of other organisms and competition for the existing resources.

A highly predictable sequence of arrivals would suggest that species' associations are important in structuring the community, indicative of an autogenic succession.

- There are three types of interaction which determine the sequence of species establishment:
  - **facilitation**: the presence or activity of a resident species enables or accelerates the establishment of a colonizing species;
  - **inhibition**: a resident species prevents or slows the establishment of a colonizing species;
  - **tolerance** is shown by species that establish late in the succession, able to survive the conditions created by the species already established.
- With facilitation a resident species modifies the habitat or provides a resource for later colonists, though it may create conditions less favourable for itself. In this case, facilitation becomes replacement.
- Inhibitory interactions can prevent some colonists establishing until much later in the sequence, or prevent them arriving at all. The sequence (and the final community) then depends on who arrives first.
- Tolerant species accommodate the conditions and the lack of resources in the later stages: they must out-compete the existing species to establish themselves. Tolerance implies better-adapted species. It therefore inhibits replacement and earlier species are lost.

Succession

A succession consists of a series of communities, or **seres**, leading to an equilibrium or climax community. In these later stages few species are able to invade and established species are unlikely to become extinct. This community is often found to:

- be dominated by long-lived and competitive species adapted to limited resources;
- have low species turnover: few colonizations and few local extinctions;
- be at its most structurally complex. The community may be stratified into distinct zones, sometimes both horizontally and vertically;
- be at its most efficient in nutrient-cycling processes, with low rates of loss.

Each of these properties arises from the established species associations.

➔ *Section 8.4, Stratification in ecosystems*

Early and late successional species

- The variable and demanding conditions on newly exposed substrates, and the travel required to find them, favours **pioneer species** that are *r*-selected.
- Tolerance among late-successional species implies many of the characteristics of *K*-selected species, when competitive interactions become important.

➔ *Section 4.2,* r-*and* K-*selected species*

Make the connection

Community structure and development and ecological processes at the species level

We may be able to predict how ecological communities are structured from some of the principles discussed in previous chapters. *Note the possible connections with each of the following topics*:

- the autecology of dominant species structuring their community: plants in many terrestrial communities, animals in some aquatic communities;
- the competitive exclusion principle and the limits on niche overlap are likely to be most important in late successional communities;
- similarly, the niche differentiation and character displacement that follows from competition and that allows co-existence;
- the niches (and resources) represented by other species may facilitate colonization;
- the adaptations of higher taxa to particular abiotic environments, especially their ability to accommodate physical and chemical gradients and the frequency of disturbance.

This list could be much longer, but consider how each might determine community structure; attempt to identify the role of these processes in the communities you have studied in detail. Consider how evolutionary processes and species interactions might be applied to the topics in this chapter.

- The dispersal and germination characteristics of its seed determine a plant's position in the successional sequence. The seeds (or propagules) of pioneer species may need to travel long distances to encounter suitable conditions for establishment.
- In temperate plant communities, mid-successional species produce berries and seeds which disperse over shorter distances, often by animals depositing the seed in their faeces. Later successional communities are typically dominated by plants with large seeds which for germination rely on animals burying caches.
- Plants have distinct **germination niches**, the conditions under which a seedling will emerge and grow. For example, seedlings in late-successional communities need to survive until conditions allow growth to maturity, perhaps when a gap in the forest canopy gives access to direct sunlight.
- Some species have a life history adapted to the frequency of disturbance in a community: **fugitive species** persist by invading gaps, where competition is less intense. They maintain a population by colonizing one gap after another, reproducing rapidly, eventually to be crowded out in each location. Fugitives contribute to the patchy mosaic of species assemblages seen in some plant communities.
- When disturbance is very frequent, **ruderals** dominate and competitive species never establish. These communities are said to 'incorporate' this frequency of disturbance and represent a **subclimax** (or, if changed by human activity, a **plagioclimax**). These persist as long as that frequency of disturbance persists. Many plant communities are held at such stages by the activity of grazers or the frequency of fire.
- The **intermediate disturbance hypothesis** suggests the highest species richness occurs at levels of disturbance which check the growth of competitive species, allowing a high species turnover (Figure 6.1).
- At climax the turnover of species will be low, with fewer arrivals and fewer departures than earlier stages.

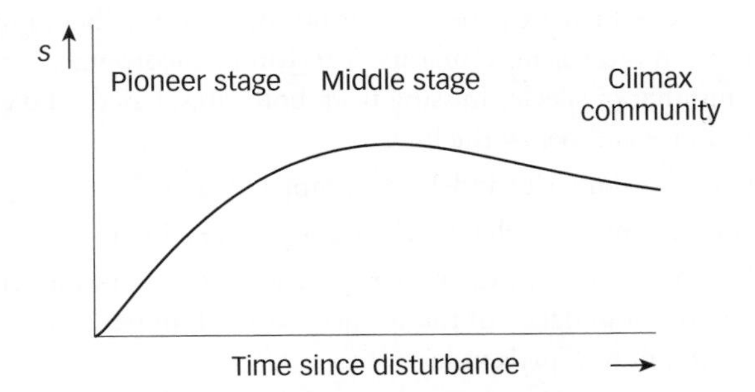

Figure 6.1 The intermediate disturbance hypothesis suggests the greatest species richness will occur in the middle stages of a succession: the number of species rises during the pioneer stage, but declines as more competitive species establish. Levels of disturbance which are not excessively disruptive can hold the community at a subclimax, allowing fewer competitive species to dominate, raising *S*.

Ruderals have traits and growth strategies to accommodate a high frequency of disturbance. Grime suggested that some plants exhibit a third strategy beyond the *r–K* continuum: a capacity to tolerate poor abiotic conditions, collectively referred to as stress. **Stress-tolerant plants** are found in soils with low nutrients, or extreme pH or metal levels, conditions which most of the competition cannot survive. Here they can afford to be slow-growing, with long generation times and low seed production.

➔ *Section 2.2, Intraspecific competition, Metal tolerant ecotypes*

Box 6.2 Key technique

Comparing communities

Repeating patterns of community composition can be an indication of their integration. Tightly integrated communities should show the same associations in equivalent habitats in different locations.

We can compare communities simply by their species lists – either as presence/absence data or by the relative abundance of the different species. It is impossible to survey entire habitats and count all species in most cases, so ecologists select an indicative group and a means of sampling it (see Box 6.3). Sampling introduces uncertainty so replication is required to measure the associated error.

➔ *Box 6.3, Classifying communities*

A variety of *coefficients of similarity* can be calculated from a species list, to compare between samples or sites. Jaccard's index is the simplest, relying on presence/absence data:

$$j = \frac{a}{a + b + c}$$

a = the number of species found at both site j and site k
b = the number of species found only at site j
c = the number of species found only at site k

Arguably the species which are expected but not present in either sample might also be important in measuring similarity. This can be incorporated by simply adding d (the number of species missing from both sites, based on data from other samples) above and below the line.

In either case, 0 = no similarity and 1 = a complete match.

Such coefficients become less reliable when sample sizes differ.

Distance coefficients measure the dissimilarity between two communities. They may incorporate the abundance of the species recorded, of which one commonly used measure is the Bray–Curtis index:

$$B = \frac{\Sigma \mid X_{ij} - X_{ik} \mid}{\Sigma\left(X_{ij} + X_{ik}\right)}$$

X_{ij} = number of individuals in species i from site j

X_{ik} = number of individuals in species i from site k

Bray–Curtis expresses the sum of the |positive| difference in the abundance of all species between two sites as a fraction of their overall abundance. Identical samples from two locations would give a distance coefficient of 0.

Samples from a large number of sites, or many samples from several sites, can then be sorted according to their similarity or dissimilarity (see Box 6.3).

There is a wide range of coefficients of similarity and dissimilarity, and different ways of sorting them. As with indices of diversity (see Box 6.4), they differ according to their sensitivity to the rare or very abundant species and to sample size. The choice of a particular measure should be informed by the habitat and the efficiency of sampling, as well as the assumptions and shortcomings of the indices chosen.

➔ *Box 6.3, Classifying communities; Box 6.4, Measuring diversity*

Explore the properties of these measures using the model in Worksheet 5 on the companion website; go to http://www.oxfordtextbooks.co.uk/orc/thrive/ or scan this image:

Box 6.3 Key technique

Classifying communities

Comparisons of communities based on their species composition will rarely find perfect matches between locations; even with the same complement of species their relative abundances are likely to differ. Ecologists need to measure their degrees of similarity and decide the limits within which a distinct community type can identified.

We would expect high similarity between samples taken within a habitat, and less so between locations. A successful classification technique will find these patterns, and measure the dissimilarity between locations. Samples can be sorted according to one of two methods.

Association analysis partitions samples by their species associations. The species recorded should be indicative and important for community structure and processes; for example, the canopy-forming trees in a woodland.

Samples in which species are commonly found together are described as 'heterogeneous'. These suggest that abiotic factors alone do not explain the distribution of at least one of the species; that is, it has an association with another species. Negative associations are found when species occur together less frequently than by chance.

With no significant positive or negative associations the samples are homogeneous and no classification is possible. Abiotic factors alone might explain the species' distributions.

continued

Significant associations allow the samples to be classified. The association is measured using a chi-square test on presence/absence data for each species in all sampling units. For each possible combination of species, a 2×2 contingency table records the number of samples containing both species, one or other species, or neither.

The classification creates a hierarchy that divides the samples progressively until only homogeneous samples remain; that is, there are no further significant associations. If the classification (and sampling) is effective, samples taken within a habitat should be grouped together consistently. Most importantly, distinct communities should separate out early in the hierarchy.

In a process opposite to the above, *cluster analysis* progressively collects similar samples together, using some measure of their difference. The result is a hierarchy in which the first groups formed have the smallest differences, and then, according to specified criteria, other sites are progressively added as a series of branches. The result is a branching hierarchy or dendrogram. The classification will depend on the properties chosen to distinguish the sites (usually several indicative species, and their relative abundance). A different dendrogram is likely to be produced if other species (indicators) are used.

The analysis uses a matrix of the distance between each sample and every other. The Bray–Curtis index is one of many possible distance measures, and nearest neighbour analysis is the simplest procedure for clustering the results: the two sites with the smallest difference become the first cluster.

➔ *Box 6.2, Comparing communities*

The next link is formed between the next two most similar sites; if this is between a third site and one of the first two, the new site is linked to the existing cluster; if not, two sites form an independent cluster. Clusters are linked in the same way. This process continues until all sites have been included in the hierarchy.

The earlier a linkage occurs in this dendrogram, the smaller the difference between the sites; sites with few species in common will only be linked further down, and are more likely to represent distinct communities.

Development of ecosystem structure and function

Ecological processes develop as the community develops. Species that occupy niches carry out roles which impart properties to the community.

Species can be collected into **guilds**, functional groups whose members carry out similar roles. For example:

- fixing energy, either as photosynthesizers or chemoautotrophs;
- releasing energy from dead and decaying organic matter;
- scavenging phosphorus;
- predating small mammals etc.

By occupying a niche a species carries out a role which imparts properties to the community.

Again, the development of ecosystem function is most easily followed in terrestrial successions.

- As a succession proceeds, most plant communities become more structurally complex, often stratified into distinct layers.
- These allow for a range of niches and opportunities for other species to utilize. The diversity of the community thus rises with an increase in its structural complexity.
- Similarly, the soil beneath becomes stratified into zones of biological activity.
- These layers have different animal and microbial communities. Particular **saprotrophs** are found at different depths down the soil profile, according to the availability of water and oxygen.
- With a supply of **detritus** from the above-ground community, the activity of the decomposer community drives the major nutrient cycles.
- The decomposer community undergoes its own successions. Organic matter arrives in pulses and each pulse induces a short-term succession in the microbial community.
- Following the reduction of its particle size by the larger invertebrates, bacteria break down the longer polymers and assimilate their nitrogen. Thereafter, fungi attack the more intractable residues.
- There is also a longer-term succession in the soil microbial community, which follows changes in the above-ground plant community. Older soils become depleted in phosphates and that reduces nitrogen fixation (Figure 6.2).
- This is one reason why stress-tolerant plant species dominate late-successional communities.
- This succession and these processes also depend on the abiotic conditions in the soil, most especially its oxygen levels, acidity, and levels of moisture.

➔ *Section 7.4, Soil nutrient cycles; Section 8.4, Stratification in soils*

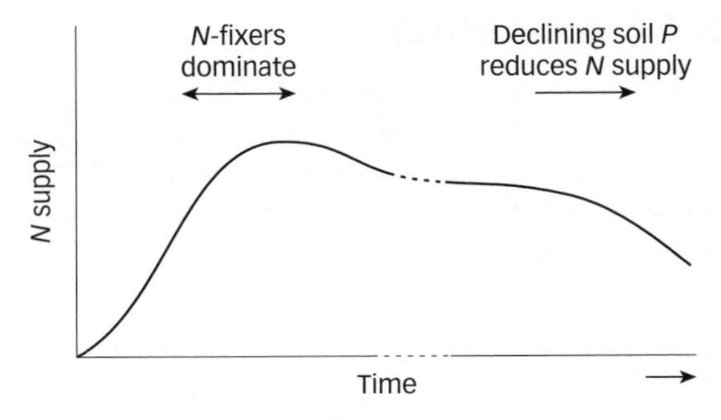

Figure 6.2 The supply of nitrogen follows the succession of saprotrophs as a soil develops in a primary succession. Bacteria able to fix nitrogen will add nitrogen soon after they colonize if there is sufficient available phosphorus. As phosphorus becomes immobilized the activity of these bacteria declines, so late-successional soils tend to be short of both nitrogen and phosphorus.

Make the connection

Increasing complexity and function as a succession proceeds

A succession helps us to see how a complex ecosystem is put together.

For example, in many terrestrial plant communities, legumes feature in the early stages of primary successions where their symbiotic associations fix nitrogen. However, as the organic content builds in the soil, phosphorus becomes less available, bound to this organic matter. Plants of the later successional stages need to form mycorrhizal associations with fungal partners that scavenge phosphorus.

These plants are long-lived and consequently tend to protect their leaves from herbivory. Their secondary metabolites, acting as anti-feedants, especially tannins and phenols, inhibit nitrifying bacteria. Thus their litter limits the availability of nitrogen in the soil. Additionally, a lack of available phosphorus also checks the growth of nitrifying bacteria (Figure 6.2).

This complex series of interactions creates the nutrient-limiting conditions typical of a late-successional soil. It also helps to explain why climax communities do not lose large quantities of these nutrients in their runoff.

➔ *Section 5.3, Plant anti-feedants; Section 5.4, Mutualistic associations; Section 7.4, Nutrient balance of late successional communities*

Notice how species composition and interactions change the ecological processes as the succession develops. Notice also how stress-tolerant plants might benefit from such effects of their litter.

Unavailable nutrients not only limit the plant community but also the activity of the decomposer community. That, in turn, limits nitrogen capture and supply to the plants. Thus the equilibrium of a late-successional plant community has its counterpart in the soil.

6.3 RULES OF ASSEMBLY

Ecologists have looked for general principles governing the structure and composition of ecological communities, to decide whether there are limits on the number of species in different roles, or certain combinations of species in different communities. Such **rules of assembly** are likely to operate in communities close to equilibrium, when species turnover is low.

However, during a succession:

- the processes of facilitation and inhibition govern the sequence in which species can be added,
- certain processes (and associated species) need to be present for other processes (and associated species) to establish,

and these also act as rules of assembly.

Beyond the predictability of an autogenic succession, are there species combinations that are unable to persist? A rule of assembly would most likely derive from species interactions.

- 'Rules' imply general constraints that structure communities and would be indicated by recurring patterns; for example, the number of species of large herbivore in grassland ecosystems.
- Rules are most likely to apply to functional groups, and the roles performed by species in certain niches.

By their presence in different locations, some species combinations are clearly more likely than others. We can also observe similar patterns across different communities, sometimes as **emergent properties** of the community itself. For example:

- late-successional plant communities are harder to invade and are dominated by long-lived species,
- early-successional communities lose a higher proportion of their nutrient inputs than late-successional communities,
- food chains tend to be limited to five or six trophic levels in most communities,

...among others.

➔ *Section 7.2, Trophic structure of communities*

Equivalent examples are found in marine and freshwater communities, and they too can have highly predictable species assemblages in some guilds.

Even if rules of assembly only apply to roles and niches, by what mechanism are they applied? Most likely from species interactions, when some part of a succession is autogenic.

- The assumption is often made that competition and niche overlap limits the number of species in a guild.
- In other cases, the effects are indirect; removing one dominant predator, for example, may cause the loss of several other species on other trophic levels.
- Some species are fundamental to the structure of the larger community. If this is a consequence of their activity rather than their abundance, these are termed **keystone species**. Their loss causes major changes to the community and its function.
- In contrast, the loss of an abundant or **dominant species** from some ecosystems (perhaps through disease or human activity) sometimes causes only negligible changes. In these cases other species have assumed the functions of the lost species.
- Certain mutualistic associations are crucial to the entire system, especially if these involve a keystone or dominant species. Then the functionality or the structure of the ecosystem changes with the loss of the association.

➔ *Section 5.4, Mutualistic associations; Section 7.2, Trophic structure of communities*

Species–area relationships

The obvious gradations in species composition along abiotic gradients suggest that allogenic factors have prime importance in some ecosystems. Similarly, in mosaic ecosystems, the turnover in species may also be driven by allogenic factors, such as the frequency of disturbance.

The consensus appears to be that:

- many species assemblages repeat according to the **replacement probabilities** of its major species; these are likely to be higher for dominant species;
- species self-organize by their associations, but also according to chance and the prevailing conditions. Again, some assemblages are more likely than others, especially in communities with keystone species;
- different combinations of species are possible as long as key roles are filled;
- convergence towards similar outcomes will follow from the selective pressures of the major abiotic factors, the history of the site, and the stability of their species configuration.

The climax at a particular site may thus be a consequence of history, chance, and circumstance.

Looking for extra marks?

Few answers

Few ecologists have attempted to write out universal rules of community assembly. But autogenic processes are known to sequence successions and equilibrium communities repeat patterns of organization. A focus on functional roles may get us closer to the mechanisms that generate these patterns.

Make the connection with the argument between monoclimax and polyclimax (continuum) models of community composition. Consider the implications of these theories for conserving habitats.

6.4 SPECIES–AREA RELATIONSHIPS

One recurring pattern is the increase in species number with sampled area (Figure 6.3). This pattern is most consistent within particular taxa.

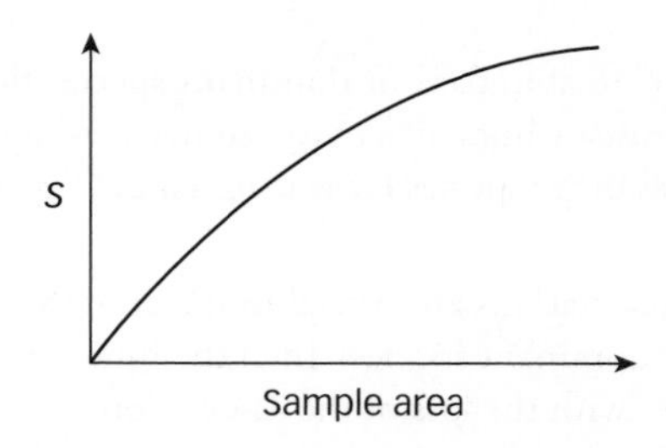

Figure 6.3 The species–area relationship. The number of species (*S*) increases with sample size, but the rate of increase slows with area sampled.

- At first the number of species increases rapidly with sampled area, but later slows. New species are found readily early on, but fewer records are added by later samples.
- This reflects the relative abundance of the different species: the common species occur in most samples and the rarities in few. Samples need to cover larger areas to find rare species with small populations.
- The produces a relationship between species number and area, and is highly predictable for particular taxa in different habitats:

  $S = cA^z$

  S = number of species
  A = area
  c = constant representing the number of species in a unit area (or sample)
  z = rate of increase of S with A, typically between 0.15 and 0.35
- z is a power term, so S increases logarithmically with area: if, for example, $z = 0.3$ a 10-fold increase in the area will only double the species count.
- z reflects differences in abundance between species: S flattens rapidly in taxa with many rare species.

The value of z also reflects the mobility of a group and the connections between the habitat patches used by these species:

- z is low in well-connected habitats because most patches have been reached by most species. Each sample is likely to share the same species so S rises slowly with area.
- For the same reason, z is low for good dispersers.
- z is high for poorly connected patches. Fewer patches are reached by all species and most patches will have a different species assemblage.
- Similarly z is high for poor dispersers.

z is an empirical term, derived from data for a group across different sampled areas, and will change as the community develops, as more species are added, or as populations change.

➔ *Section 8.3, Connectivity in landscape ecology*

Island-biogeography theory

Islands are habitats isolated by conditions hostile to their inhabitants. Well-defined boundaries allow us to measure the colonization and extinction of a habitat over time.

MacArthur and Wilson's model of island biogeography suggests the number of species is a balance between rates of colonization and extinction. Although most often discussed in terms of oceanic islands, the model also applies to habitat islands, isolated by unfavourable conditions for the species of interest; for example, fish in freshwater ponds scattered across an agricultural landscape. Again, the model works best for particular taxa or guilds.

- The model does not predict which species will form the stable community, only the number (*S*) at equilibrium, when colonizations are balanced by extinctions.
- This is a dynamic equilibrium: *S* remains the same but the composition changes as colonization and extinction proceed (Figure 6.4a).
- *S* is fixed by island size so the model uses area as a measure of limiting resources and available niches.
- Initially niches and resources are unoccupied and new arrivals establish easily. Later, when most are filled, rates of colonization decline.
- Then competitive interactions make extinction more likely and colonization less likely. Since the intensity of competition increases with *S* neither rate is expected to be linear (Figure 6.4).
- In a process comparable to succession, those able to make best use of the ecological space are most likely to persist.

Islands differ not only in size but also in proximity to the source of their colonizers.

- An island close to the mainland (or source community) will acquire species more rapidly and have a higher *S*.
- Populations on more distant islands suffer higher extinctions because fewer individuals arrive to supplement their numbers.
- Smaller islands supporting smaller populations will also have higher extinction rates. A small island far away has a small *S*.
- We can thus identify four possible equilibria for *S*, between near and far, and large and small islands (Figure 6.4b).

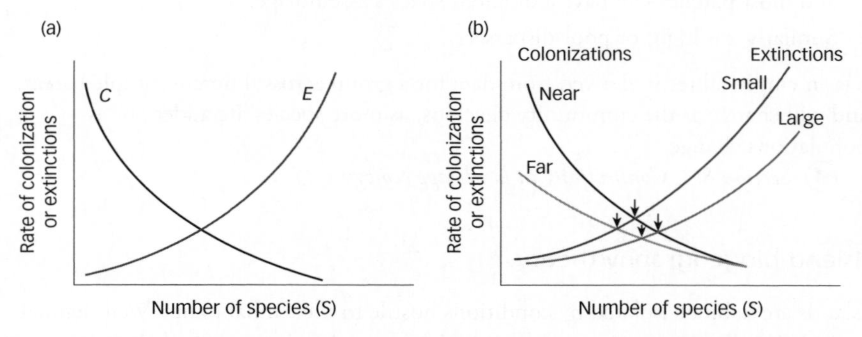

Figure 6.4 The MacArthur–Wilson theory of island biogeography predicts the equilibrium number of species of a particular group (such as birds or reptiles) on an island. This is a dynamic equilibrium, the balance between local extinctions and new arrivals.

(a) For any island, rates of colonization *(C)* decline as species accumulate and rates of extinction *(E)* rise as niches are occupied. Their intersection gives the equilibrium number of species.

(b) *C* changes with island proximity and *E* with island size. The value of *S* for four combinations of size and isolation are indicated by the arrows.

The model has some success in predicting *S* within island and species groups. However, it omits detail likely to be important in determining *S*.

- Species fill different niches and are not equivalent: some facilitate and others inhibit later colonizers.
- Over ecological time, new species are likely to evolve. The intensity of competition, especially when resources are scarce, will promote niche differentiation.
- Dispersal and the frequency of genetic exchange between habitats will determine the isolation of their gene pools and local speciation.
- Founder effects as well as genetic drift promote speciation in isolated populations.
- For these reasons a high proportion of endemic species is a feature of many oceanic islands and isolated habitats. Islands also tend to have fewer species than equivalent mainland areas.

➔ *Section 3.3, Allopatric speciation*

More elaborate models allow for speciation, both in the source community and on the island. Others recognize that the most abundant species on the mainland are most likely to appear on the islands. Some models incorporate measures of productivity for the islands as an indicator of available resources, and of community complexity and the potential range of ecological niches.

For some islands, colonization represents a primary succession, for others it is a secondary succession. Like all ecosystems, the turnover of species means that its assemblage changes through time, both through changes in the abiotic and biotic environment, and with niche differentiation, extinctions, and colonizations.

➔ *Section 3.3, Allopatric speciation; Section 8.5, Global gradients in diversity*

Looking for extra marks?

See whether you can find an example of an archipelago (other than the Galapagos!), or some other sequence of 'island' habitats, where you can relate the number of species (of a particular taxon or guild) and their endemism to 'island' size and isolation.

Species richness

As with island ecosystems, the number of species in a habitat patch should increase during a succession towards an equilibrium, when turnover is minimized.

S, **species richness**, is the number of species in a community.

- This has to be distinguished from **species diversity**, which combines species richness with a measure of the relative abundance of the different species: **species equitability** (Box 6.4).
- At equilibrium, *S* should be equivalent to the number of available niches, and equitability is a measure of the available resources in each niche (indicated by the abundance of its occupant).

Box 6.4 Key technique

Measuring diversity

Species diversity consists of two components: the number of species (species richness) and the distribution of individuals between these species (species equitability; see Figure 6.5a). A diverse community will have a high species richness and an equitable distribution of individuals between those species. Low diversity can be due to either low species richness (few species) or low equitability (most individuals represented by few species), or both.

Of the many different measures, indices of diversity can be divided into two broad types: distribution-dependent and distribution-free. Distribution-dependent indices assume the relative abundance of different species approximates to a particular pattern, of which several have been proposed. Most have been justified according to models of resource partitioning and niche space. However, to be effective these indices require large and efficient sampling programmes (as implied by the species–area relationship).

➔ *Section 6.4, Species–area relationships*

Distribution-free indices make no assumptions about relative abundance, and have been used far more frequently.

Simpson's index (D) measures the probability of collecting two individuals of the same species consecutively when sampling a community. Where there is high species richness and equitability, the chances of doing this will be small.

If there are n_i individuals of species i, the chance of selecting one i from the total of N individuals in the sample is:

$$\frac{n_i}{N}$$

Having removed this individual, the chances of doing this a second time becomes

$$\frac{n_i(n_i - 1)}{N(N - 1)}$$

To obtain the index, we carry out this calculation for every species in our sample, add them together and subtract the result from one:

$$D = 1 - \Sigma\left[n_i(n_i - 1)/N(N - 1)\right]$$

D ranges from 0 to 1, from low to high diversity. The main drawback of D is its insensitivity to small samples.

Another commonly used index is the Shannon–Wiener information statistic, H'. This measures 'uncertainty'—how difficult it is to predict the species represented by the next individual collected—again, with high S and high equitability this uncertainty increases.

$$H' = \log N - 1/N \ \Sigma \ (n_i \log n_i)$$

Here the maximum value of the index is set by log N, the total number of individuals. The index can range from below 1 to log N, so there is no simple scale against which to compare it.

However, we can express H' as a proportion of the maximum possible value of the sample at its highest equitability: the value of H' is computed again but now with N divided equally between all species to give H_{max}.

We then derive:

$$E = H'/H_{max}$$

E ranges from 0 (no diversity, one species only present) to 1 (maximum species equitability). E is the proportionate equitability for that sample only, so limiting comparisons between samples with large differences in S.

All indices of diversity depend on the size of the sample and its efficiency, and how closely it represents the larger community. Most work best with particular groups of species. With consistent sampling over time or between locations they can be used to follow community development and response to disturbance.

Example calculations are available in Worksheet 6 on the companion website; go to http://www.oxfordtextbooks.co.uk/orc/thrive/ or scan this image:

More than one type of **diversity** is used to compare ecosystems.

- Within a habitat the number of niches is a measure of **habitat diversity**. In practice, this is difficult to measure: most often it is equated with the number of resident species!
- At a higher level, habitat diversity can be measured in a landscape or a region, as the number and proportion of different habitat or patch types.
- **Genetic diversity** is the number of genotypes within a population. This depends, of course, on the loci selected for measurement, but will be larger in larger populations and larger still between ecotypes.

All of these are likely to increase with sampled area.

Species diversity can be partitioned in different ways:

- α diversity is measured within a habitat,
- β diversity is the change in diversity between habitats.

If measured as species richness:

- the β diversity between two habitats with no shared species would be the sum of their S,
- otherwise it is the sum of the number of species unique to each habitat.

Species–area relationships

γ diversity is the overall diversity for a region or landscape. This might be measured as the number of species recorded for a landscape.

➔ *Section 8.3, Landscape ecology*

The simplest measure of diversity is the *S*/*N* ratio—the number of species divided by the total number of individuals—but this has limited utility. However, even elaborate indices of species diversity can fail to properly represent the true diversity (Figure 6.5).

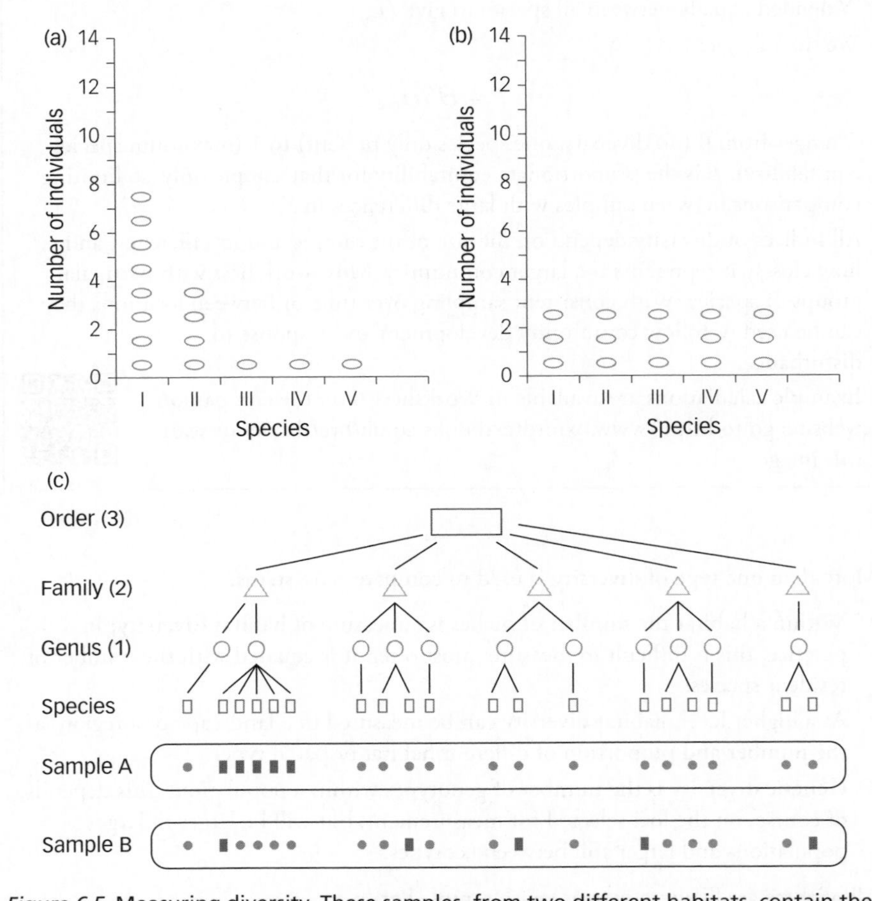

Figure 6.5 Measuring diversity. These samples, from two different habitats, contain the same number of species ($S = 5$) and the same total number of individuals ($N = 15$). They differ in their species equitability, which the *S*/*N* ratio fails to measure. Community (a), dominated by just two species, is less diverse than (b) with its high equitability.

Most measures of diversity fail to score the *range* of species found in a habitat because they do not measure the phylogenetic spread of its species. This can be done by including taxonomic distance: the number of taxonomic steps between individuals (c).

Steps are weighted (values in parentheses) so species separating higher up the hierarchy score more; for example, two individuals from different families have a taxonomic distance of 3 (genus (1) + family (2)). Sample B has five individuals from five different families, giving it an average taxonomic distance three times that of sample A, with its five species from the same genus.

Species diversity tends to increase as a succession proceeds and as ecosystem function develops. Indeed, conservation strategies invariably seek to protect or restore these functions to sustain its diversity. This prompts the question of which species are essential to maintain ecosystem function.

Functional redundancy and the stability of ecosystems

Beyond the dominant species, are there others which, if lost, would radically change the community? Or, which species could be lost with no change in ecological processes?

- The capacity to maintain an ecological process when a species is lost is termed **functional redundancy**. A high functional redundancy is likely to be a feature of resilient communities.

➔ *Section 7.3, Ecosystem stability*

- For example, niche differentiation between certain soil microorganisms appears slight, and many seem to do equivalent jobs. Most soil processes are highly resilient to disturbance and species loss, suggesting that surviving species within a guild are able to maintain the function (Figure 6.6).
- The best evidence for functional redundancy comes from experiments that manipulate soil communities. It also appears true of many highly diverse communities, including coral reefs.
- This redundancy, and the replication of roles within guilds, may explain why some highly diverse communities recover rapidly from major disturbances.

➔ *Section 8.1, Functional redundancy in coral reefs*

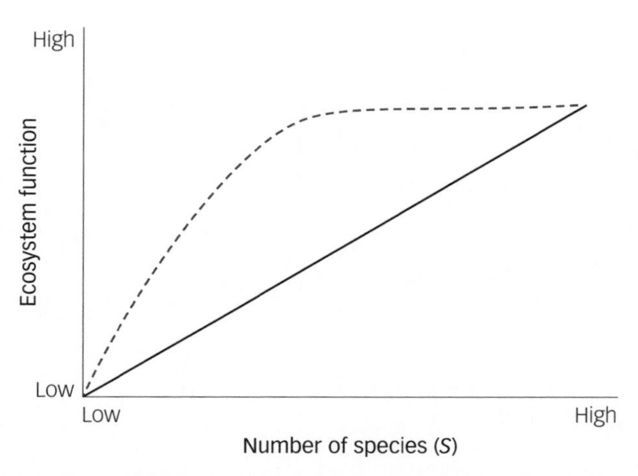

Figure 6.6 Functional redundancy in ecological communities. If every species is equally important to an ecological process we expect a linear decline in function as species are lost (the solid line). If, instead, some species are redundant the ecological function will not decline linearly (the broken line).

Species–area relationships

Tropical forests, among the oldest terrestrial ecosystems on Earth, are also the most diverse. They are home to some of the most ancient phyla, and have persisted, in places, through major climatic upheavals. One possibility is that their resilience is a function of their diversity.

➔ *Section 8.2, Global gradients in diversity*

Check your understanding

Examination-type questions

1. Why should the monoclimax model of succession imply that communities are structured by their species interactions? Illustrate your answer with examples.
2. At the level of the landscape, explain how the intermediate disturbance hypothesis could explain the increased species richness of a plant community composed of a mosaic of local patches cycling through different stages.

You'll find answers to these questions—plus additional exercises and multiple-choice questions—in the Online Resource Centre accompanying this revision guide. Go to http://www.oxfordtextbooks.co.uk/orc/thrive or scan this image:

7 The ecology of the ecosystem

Key concepts

- The ecological community together with its abiotic environment comprise an ecosystem. Systems ecology models the structure, dynamics, and properties of ecosystems.
- Compartment models simplify complex feeding relations within an ecosystem and quantify the transfer of energy between trophic levels.
- Energy is fixed in primary production from sunlight or from inorganic chemical compounds. There are three types of photosynthesis in higher plants.
- The size of successive trophic levels and the number of links in a food chain follow from the inefficiency of energy transfers.
- The complexity of food webs increases as the community develops. Complex ecosystems were thought to be stable, although modelling suggests their resilience and elasticity depends on guilds with high redundancy.
- The flux of a nutrient through an ecosystem depends on its chemistry and role in biological systems.
- Nutrient and energy transfers are closely linked: the movement of nitrogen through most terrestrial ecosystems is governed by energy capture and availability to the decomposer community.

continued

- Carbon compounds are the energy currency of the biosphere. The residence time and concentration of carbon are major determinants of atmospheric mean temperature.

Assumed knowledge

The basic biochemistry of metabolism and photosynthesis. Simple energetics and the laws of thermodynamics. The hydrological cycle.

A conceptual map for this chapter is available on the companion website; go to http://www.oxfordtextbooks.co.uk/orc/thrive/ or scan this image:

7.1 MODELLING ECOSYSTEM PROCESSES

An ecological community together with its abiotic environment comprises an ecosystem. The abiotic environment will change as a community develops. For example, a soil becomes stratified as biological activity and organic matter accumulate in its upper layers; the growth of its polyps extend the scaffold provided by a coral reef and the habitats for other species.

We can follow an ecosystem's development by measuring its properties and processes, such as the rates at which energy or nutrients move through the system.

Systems ecology

Most natural ecosystems have too many variables for experimentation under controlled conditions. We have to observe their processes and attempt to correlate these with likely regulatory factors.

As an alternative, ecologists create smaller versions to replicate and control conditions:

- **microcosms** are artificial ecosystems, reduced in size and complexity, constructed to follow a small number of processes. Often laboratory-based, they do not attempt to mimic the complexity of the larger system;
- **mesocosms** partition real ecosystems to create replicates, or define boundaries across which we can monitor inputs and outputs, and allow experimental manipulation. Lakes and watersheds have been used in this way.

Revision tip

In your review of any microcosm or mesocosm studies, ensure you understand the purpose of the experiment and the extent to which the published data answer the hypothesis.

How would you improve the method? What other data might be worth collecting?

Following a natural ecosystem over time requires a large effort, not only in collecting data but in describing its organization and deciding which measurements are needed. It requires a statement of purpose—a hypothesis to be tested—and some metric to provide the answer.

With an accurate description of the system, ecologists can model its properties. Such indicators might include (among others):

- species richness and relative abundance,
- structural complexity,
- spatial extent,
- longevity,
- rates of production,
- rates of decomposition,
- rates of nutrient capture and transfer,
- process stability and capacity to resist change.

➔ *Box 6.4, Measuring diversity*

Systems ecology treats an ecosystem rather like a machine, describing its important components and the factors regulating its functions. This recognizes that all ecosystems are driven by energy, and that:

- to persist, rates of production need to exceed rates of decomposition;
- energy capture and release drive nutrients through the system;
- the strength of a disturbance can be measured by changes in these rates;
- regulation is achieved by negative feedback;
- these properties are an indication of how the ecosystem is organized.

Note the assumption that the system has an organization that is conserved.

Systems ecology creates a model and measures key inputs and outputs, according to the process and hypothesis being tested. This is often the only practical means of understanding large and complex natural systems.

- Models divide an ecosystem into compartments (subsystems) and describe the relations between these compartments. How an ecosystem is partitioned is critical for the purpose of the model, and for the measurements taken (Figure 7.1).
- An effective model will accurately predict an ecosystem's response to a disturbance. Once validated it can serve to test hypotheses and establish causal links (Box 7.1).

Based on their purpose, models are of two basic types.

- *Analytical models* simplify the system to isolate its controlling factors for some process. They are deterministic: a particular input allows for just one possible output (or a limited range).

Some analytical models are stochastic—they incorporate biological variability—so one input can give a range of outputs. Analytical models are often the first steps in understanding a complex system.

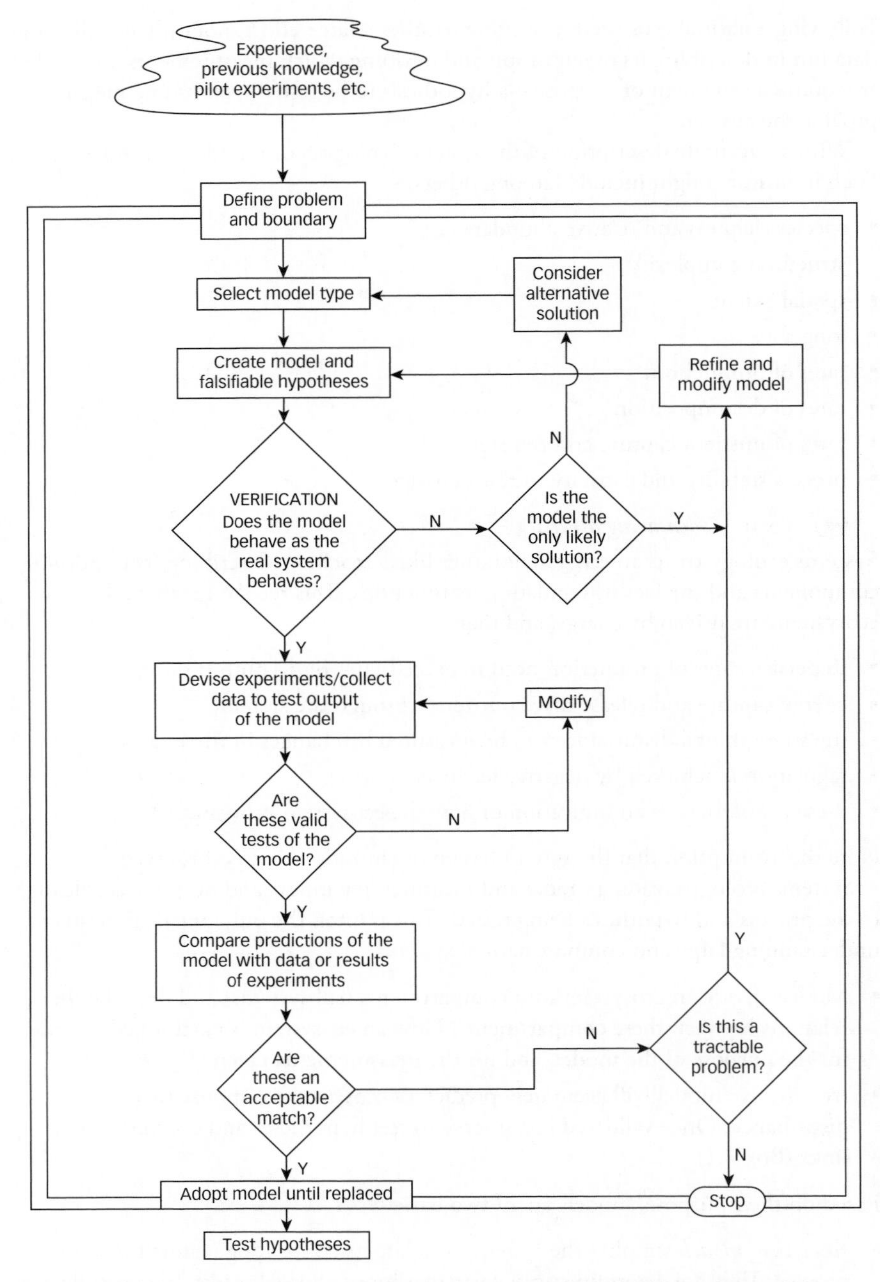

Figure 7.1 A flow diagram of the steps in the creation and validation of a model. A model can be considered as a series of linked hypotheses, each of which must be testable and falsifiable. A model is validated when it follows the general behaviour of the system over a wide range of conditions, perhaps including experimental manipulations. Models that survive a large number of such tests—whose outputs closely match observations—inspire high confidence. Otherwise, the model has to be refined to improve its predictive performance.

Box 7.1 Key technique

Principles of modelling

Most modelling is an exercise in computer programming and can be visualized as a flow chart (Figure 7.1): the system is partitioned into a series of compartments and the relationships between them are quantified.

Just like a computer program, the model is designed for a purpose and its boundaries and compartmentalization reflect this: it would change if a different process or some other output were measured. Simple dynamic models have four elements, as follows:

i. *System definition*: a system consists of two or more interacting components and a boundary that delimits the area of interest. Material or information can cross this boundary as inputs or outputs. Outputs can be used as indicators of the performance of the system; for example, a simple material balance, inputs minus outputs, will indicate whether the system is accumulating or losing a nutrient.

ii. *Variables*
   a. External variables (or forcing functions) are the external factors operating on the system; for example, ambient temperature.
   b. State variables are the components of the system. They indicate its state by their level or value; for example, the percentage of organic matter in the soil.

iii. *Processes*: these are the equations that describe the relationships between the state variables; for example, the rate at which organic matter is decomposed in the soil. Rates may vary with the size of the state variable. If there is negative feedback the system is regulated.

iv. *Parameters*: these are constants that describe fixed relationships between variables; for example, the water-holding capacity of the soil given its percentage organic matter content.

In Figure 7.1 we model the process of modelling, showing the steps and decisions in the creation and validation of a model. This illustrates the conventions used in flow charts and shows the stepwise decomposition of the system required to create the model.

A flow chart helps us to review the model and isolate the source of inaccuracies. Sensitivity analysis can identify those elements most important for determining the outcome: the variables or processes responsible for large changes in the outputs.

More elaborate models allow for variation in the rate processes and state variables. Using matrix algebra or multivariate techniques, detailed simulations can incorporate this stochasticity, generating a range of possible outcomes and their probabilities, rather like modern weather forecasts.

The population models of Section 4.4 are deterministic, where the output is the change in population size. Here only the birth and death rates decide the rate of population growth.

➔ *Section 4.4, Models of population growth*

- *Simulation models* include the detail required to capture some of the system's complexity. The emphasis is on description, to mimic the response of an ecosystem over a wide range of conditions. They are usually stochastic and require large amounts of observational data, collected over a range of conditions.

Rather like a weather forecast, an ecological model is judged by its predictive power.

7.2 ECOLOGICAL ENERGETICS

An organism is a device for storing energy in various chemical bonds. Life is a series of mechanisms for capturing energy to make these bonds and releasing it on demand.

- Energy is the capacity to do work; metabolism is the collection of biochemical pathways that store and release energy.
- Metabolism uses energy to create molecules (*anabolism*) and to degrade them (*catabolism*).
- Anabolism creates energy stores and structural molecules. Catabolism is used to release energy and recover materials.

Individuals use energy to drive their growth and reproduction (Figure 2.1), but they are also a potential source of energy for other organisms. Ecosystems develop (and persist) because other species exploit these sources, whether it is the energy stored in the tissues or lost with their waste (Figure 7.2).

Types of nutrition

Autotrophs use ambient energy (either radiant or chemical) to create energy-rich carbon compounds, using an inorganic source of carbon. Thus carbon dioxide is reduced by combining with hydrogen through photosynthesis. **Heterotrophs** release the energy from carbon compounds formed by other organisms.

Autotrophs are described as **primary producers**.

Energy fixation

All primary producers combine hydrogen with a carbon molecule to create an energy-rich compound, but use different sources of carbon and hydrogen. They also use different sources of energy. This energy is fixed in one of three ways.

- **Photoautotrophs** use sunlight and inorganic carbon (carbon dioxide). Besides the higher plants, this group includes algae, cyanobacteria, and purple sulfur bacteria. Most use water as a source of hydrogen (and electrons) but the last group use hydrogen sulfide.
- **Photoheterotrophs** use sunlight and organic carbon. These are the non-sulfur purple bacteria.

- **Chemoautotrophs** oxidize inorganic compounds as their energy source and carbon dioxide as their carbon source. The bacteria in this group can be classified by the compounds they oxidize and include the nitrifying bacteria.

The rest—animals, fungi, protozoa, and other bacteria—rely on energy that has already been fixed in organic compounds:

- **Chemoheterotrophs** use organic compounds as both energy and carbon sources. **Saprotrophs** feed on dead or decaying organic matter, *biotrophs* feed on living organisms.

All cellular life relies on coupled oxidation-reduction (redox) reactions to drive their metabolism. This involves the transfer of one or more electrons between a donor compound and an acceptor.

Reminder

Redox reactions and respiration reviewed

Oxidation is the loss of electrons	Reduction is the gain of electrons
Reducing agents lose electrons	Oxidizing agents gain electrons

Molecules gain energy when they gain electrons

Respiration gives up energy.

Aerobic respiration is a redox reaction: oxygen is the oxidizing agent and glucose is the reducing agent.

$$C_6H_{12}O_6 + 6O_2 \rightarrow 6CO_2 + 6H_2O + \text{energy}$$

Hydrogen atoms (and their electrons) are donated by glucose and combine with oxygen (the electron acceptor) to form water. Since oxygen is highly electronegative, water has lower-energy bonds than glucose, and the reaction therefore releases energy. This is used to produce ATP.

Energy release can take various forms, especially among the bacteria. These reactions invariably use dehydrogenation: the transfer of electrons (and of hydrogen) from the donor molecules, carbon compounds, to an electron acceptor. There are three types of reaction by which energy is so released.

- *Respiration* uses molecular oxygen as the final electron acceptor: oxygen is the oxidizing agent. Carbon dioxide and water are by-products.
- *Anaerobic respiration* uses an inorganic oxidizing agent—principally sulfates, nitrates, or carbonates. The by-product is, for example, hydrogen sulfide.
- *Fermentation* uses an organic compound as the oxidizing agent but no oxygen is used (so no water is produced). It is therefore another form of anaerobic respiration. A range of by-products are possible, including alcohol and lactic acid.

Higher photoautotrophs use only aerobic respiration, although photosynthetic bacteria are all anaerobes.

- Most organisms are not constrained to one form of energy release. The greatest range is found in the bacteria, just as they have the greatest range of energy-capture methods.

- Some can only use one metabolism (e.g. obligate anaerobes) but others are able to switch as the environmental conditions change (e.g. facultative anaerobes).
- The metabolic range of bacteria reflects their ancient phylogeny. It explains their capacity to survive extreme environments and the range of inorganic compounds they exploit as energy sources.
- It also explains their vast diversity in any ecosystem and the fine niche partitioning seen in the microbial decomposition of even uniform substrates, such as wood.

Looking for extra marks?

Fine niche partitioning is most evident in microbial systems

It may be worth considering less obvious ecosystems to find examples of rapid character displacement and niche differentiation, especially among bacteria and fungi. *Consider the range of life in a rotting log, a yogurt, or a digestive system.*

Artificial ecosystems—microcosms such as fermentors—allow us to isolate the strains and varieties that flourish under particular biotic and abiotic conditions.

Types of photosynthesis

Disregarding energy imported from other ecosystems, the proportion of received energy fixed in a terrestrial community is set by the photosynthetic efficiency of its primary producers. These efficiencies are low (0.5—2%), in part because energy is captured from just the red and violet-blue part of the visible spectrum. These rates vary between plant groups.

Opening the stomata to allow carbon dioxide into the leaf allows water to escape, so the availability of water is often a limiting factor.

- Shade-tolerant plants have lower rates of photosynthesis and lower rates of metabolism. They grow beneath the leaves of other plants and make trade-offs to survive the lower light intensities, most especially in their lower drought tolerance.
- Plant groups also differ in the biochemistry of their photosynthesis and the configuration of their chloroplasts.
- Most plants produce a three-carbon acid from carbon dioxide as the first stage in carbon capture. These are termed C_3 plants.
- A smaller number produce four-carbon acids, using an additional enzyme less sensitive to the presence of oxygen. They also have their chloroplasts arranged in two layers around their leaf veins, with a high concentration of mitochondria.
- Because they bind carbon dioxide at lower concentrations, these C_4 plants are more efficient photosynthesizers. They fix carbon dioxide even with their stomata closed, especially at high light intensities and high temperatures when water loss would otherwise be high.
- For this reason, C_4 grasses are more common than C_3 grasses in hotter and drier areas, especially in the tropics.
- Many desert plants only open their stomata at night. Carbon dioxide fixed as organic acids is later released to drive photosynthesis during the day, when the stomata are closed. This is crassulacean acid metabolism (CAM), found in

succulents and true cacti, plants which store water to support photosynthesis when the stomata are closed. They have various other anatomical adaptations to reduce water loss.
- Photosynthesis first evolved in bacteria, using chlorophyll held in sacs beneath the cell membrane. Their reaction is highly sensitive to oxygen and does not use water. Instead, sulfur, sulfides, or hydrogen are electron donors.

Presumably the greater efficiency of C_4 plants involves some trade-off that prevents them displacing C_3 plants more widely.

Looking for extra marks?

Match the type of photosynthesis to the ecosystem

These different forms of photosynthesis can be correlated with particular habitat conditions. You are likely to score highly if your description includes the anatomical adaptations seen in higher plants, both to conserve water and make efficient use of carbon dioxide.

For example, what are the advantages for C_4 plants in concentrating their chloroplasts in bundle sheaths? What selective advantage do C_3 plants enjoy? Why is the concentration of oxygen so critical to the function of chloroplasts?

You may be required to describe the biochemistry of these different metabolisms in some detail.

Measuring the energetics of an ecosystem

Ensure you know the main measures of energy fixation and transfer in an ecosystem and can derive the efficiencies when presented with a set of data (Box 7.2).

Ecologists distinguish between **primary production** (the energy fixed by autotrophs) and **secondary production** (the energy fixed in the tissues of consumers, including detritivores).

The methods used to measure community energetics vary with the organisms and the ecosystem. Aquatic systems are sometimes measured by volume, terrestrial ecosystems (other than soils) by area.

You should know the sampling techniques in ecosystems you have studied.

- **Biomass** (or standing crop) is the mass of organisms (or its energy equivalent) in a given area or volume, at a particular time (e.g. g/m^2 or kJ/m^2).
- This can be measured for an entire ecosystem or for a particular compartment (a trophic level, guild, or species).
- It should specify which components are being measured (living biomass only, living + dead tissues) and their state (fresh, dry, or ash weight).
- Measuring the addition of new biomass over time gives the rate of production (as $g/m^2/year$ or $kJ/m^2/year$).
- This allows calculation of the **turnover time** (transit time or residence time) for a quantum of production, simply the standing crop divided by the production rate. This is the time required to replace the standing crop.

Box 7.2 Measures of primary and secondary productivity and ecological efficiencies

Production (P)

Biomass or *standing crop* is the production accumulated per unit area or volume at a particular time.
Note: biomass can be mass or energy per unit area or volume.
Turnover time is the rate of replacement of a unit of biomass, the time required to replace the standing crop:

$$= \text{biomass/net productivity rate} = \frac{\text{g/m}^2}{\text{g/m}^2\text{/year}}$$

For primary producers

Gross primary production (GPP)
GPP = energy fixed/unit area or volume/unit time
Net primary production (NPP)
NPP = GPP − energy lost with respiration (R)/unit time
Photosynthetic (Ps) efficiency
Ps efficiency = NPP/radiant energy received
Compensation point
GPP = R when there is no NPP

For individual consumers

Consumption efficiency (*CE*)
CE = energy consumed (*C*)/energy available in diet
Assimilation efficiency (*AE*)
AE = energy assimilated (*A*)/energy consumed (*C*)

AE is the proportion of energy taken into the tissues, before losses to respiration ($A = P + R$)

Production efficiency (*PE*)
PE = energy fixed in tissues (*P*)/energy assimilated (*A*) or
energy fixed in tissues (*P*)/energy consumed (*C*)

PE is the energy fixed in the tissues after respiration:
P/*A* is *PE* expressed as a proportion of the energy assimilated
P/*C* is *PE* expressed as a proportion of the energy consumed (before faecal losses).

P/*A* differs between animal groups: homoiotherms have a low *PE* because of the high costs of their respiration.
Poikilotherms are more efficient. Consequently a larger proportion of their assimilated energy is available to consumers of their tissues.

Between trophic levels

Trophic level production efficiency (TLE)
$\text{TLE} = P_{x+1}/P_x$
between trophic level x and $x + 1$.

P here represents the production for all species on a trophic level. When calculated for the entire trophic level $CE \times AE \times P/A$ is equivalent to TLE.
Note these efficiencies vary between trophic levels and especially between ecosystems.

- Measured for primary or secondary production, the turnover time gives the speed at which the system, or a compartment, is running. Note this can be independent of size of the standing crop.
- Turnover tends to be slower where nutrients are limiting, but faster if there is frequent disturbance (or grazing), releasing nutrients. A large standing crop can be accumulated over a long period with a slow turnover rate, typical of many late successional communities.

➔ *Section 6.2, Ecosystem function and succession*

Looking for extra marks?

Indicators of ecosystem processes measure community change

Make the connections between the low turnover rates of many late-successional and nutrient-limited ecosystems with their highly competitive communities, dominated by species with long generation times. Turnover times indicate ecosystem function and reflect changes in the species assemblage with disturbance.

They also reflect the differences between ecosystems, and their abiotic conditions. Contrast, for example, the standing crop and turnover times between temperate and tropical forests.

Secondary production

Energy moving from primary producers to herbivores and higher trophic levels represents a **food chain**. Ecosystems have two food chains: one supported by primary production, the **grazing food chain**, and one supported by dead and decaying organic matter, the **decomposer food chain** (Figure 7.2).

A relatively small proportion of the energy entering one trophic level actually passes to the next trophic level. Energy is lost as heat from organisms due to their respiration, and with their waste.

- The rate of respiratory loss differs between animal groups according to their physiology, especially between **poikilotherms** and **homoiotherms**.
- Resting or basal metabolic rate varies with body size and life history strategy.

  ➔ *Section 2.1, Acclimation and basal metabolic rate*
- Consumers do not assimilate all the energy they consume and a significant proportion is lost, undigested, with their faecal waste.
- Many organisms lose energy through deciduous tissues (leaves, hair, **exuviae**, and so on), further reducing that available to the trophic level above.
- Energy is also lost when individuals or organic matter leave an ecosystem or are prevented from decomposing.

Some organisms are not consumed by the trophic level above, but die and pass to the other side, entering the decomposer food chain.

- In terrestrial plant communities, most of the primary production passes only through the decomposer food chain. A large proportion of this **detritus** is consumed by saprotrophs.

Ecological energetics

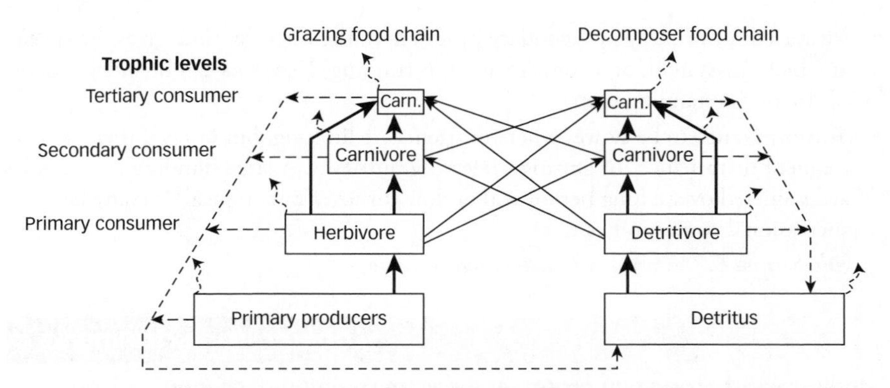

Figure 7.2 A generalized model of energy flow through an ecosystem. This includes the energy exchanges that can occur between the grazing and decomposer food chains, but omits microbial pathways and the routes associated with omnivore pathways and parasites: a simplification made in most studies of energy budgets.

Energy following lines of consumption are shown by the solid lines; energy passing to the decomposer food chain is shown by the hatched lines; energy lost through respiration by the dotted arrows.

If, instead of these trophic levels, individual species were represented, this would be a model of a food web.

- Most of the energy in the grazing food chain will eventually enter the decomposer chain as waste or dead tissues.
- Energy can also move between chains by carnivores feeding in both chains.
- Energy cannot be re-used, although it may cycle (unused) through the decomposer food chain several times.

A food chain simplifies the ecosystem and creates a compartment model for measuring transfer efficiencies. The low rate of energy transfers can be represented as pyramids of declining energy or biomass in successive trophic levels.

However, the fullest description of these pathways is a *food web*.

Ensure that you can attribute each of the routes (arrows) mapped in Figure 7.2.

Looking for extra marks?

Succession and classification in decomposition

The succession of species degrading plant material reflects the costs and returns from attacking this substrate. After fragmentation by the large soil invertebrates (earthworms, woodlice, mites, and others), decomposition follows a sequence where the easiest molecules to utilize (the simplest) are consumed first. After these sugars, the more intractable remains are attacked: the long-chained and complex carbohydrates which incur greater costs and require a larger range of enzymes. Some residues, such as lignin—and later the humic and fulvic acids—can only be degraded by specialist bacteria and fungi.

Many of these bacterial species are differentiated phylogenetically by their metabolisms and favoured substrates.

Chains, pyramids, and webs

Food chains represent the transfer of energy through a community or along a specific pathway.

- Within an ecosystem the relative size of sequential trophic levels determine the length of a food chain and are indicative of the standing crop at each level.
- The similarity of food chains between ecosystems suggests that community structure is often limited by its energy transfers:
  - few grazing chains extend beyond five or six links;
  - the decrease in biomass or numbers with higher trophic levels suggests that either energy or nutrients are limiting for top consumers;
  - larger animals are found at the end of the chain, with longer lifespans, smaller populations, and lower reproductive rates.

  ➔ *Section 6.3, Rules of assembly*

Make the connection

Note these limitations on available resources—energy and nutrients—set the carrying capacity for larger organisms and determine their reproductive strategy. It is the struggle for these limited resources that favours competitive species at these higher trophic levels.

- Following a single pathway—just one consumer and its predator, for example—can give precise transfer efficiencies, especially with replicated observations. The practicalities of collecting these data depend on the prey having a small number of predators and a restricted diet.
- However, there can be problems of interpretation, not least in allocating species to a compartment:
  - many feeding relations are not linear, and omnivorous scavengers may feed on trophic levels above and below them, and between food chains;
  - feeding relations change with time and abundance. Our methods do not always allow for these changes, or for seasonal migrations and occasional visitors;
  - should the energy represented by the parasites hosted by a herbivore count towards the primary consumer level...or another one? Does this depend on whether the parasites form part of the diet of the carnivore...or go on to infect it?
- We can also follow the transfer of key nutrients between trophic levels and compare turnover times as a community develops.

Looking for extra marks?

Viable feeding strategies

The inefficiency of the transfers can explain the life history strategies of consumers on each trophic level: why a ruminant feeds continuously, or a carnivore occasionally; why there are so many wildebeest on the African savanna and relatively few lions.

Ensure that you can illustrate the autoecology of one or two species with reference to their trophic position. For example, why should deciduous trees shed their leaves during the winter and nearly all conifers retain them? Or why should specializing on one prey species, for example a parasitoid, be a viable strategy compared to generalized insect predators?

Useful insights often come from comparing closely related species: how do the feeding habits of a wolf and a fox differ, and reflect their respective niches?

Trophic pyramids of energy, biomass, or abundance allow comparisons between trophic levels and also between ecosystems. However, they can be a poor representation of the system's energetics:

- Decomposer communities may not show declining biomass or abundance at successive levels.
- Some ecosystems are constructed on a single primary producer—an oak tree, for example—and a pyramid of numbers therefore looks precarious.
- Pyramids do not capture the recycling of nutrients crucial to the organization of ecosystems.
- The pattern assumes a constant supply of resources yet some communities are transitory: the secondary productivity of freshwater streams in temperate regions depend on an autumnal pulse of **allocthonous** organic matter from the surrounding terrestrial vegetation.
- The trophic pyramid is often used to explain the transfer of pollutants and their increasing concentration at higher trophic levels, but this is far from universal. For example, cadmium shows **biomagnification** along terrestrial grazing chains whereas lead rarely does.

A **food web** is a map of the energy pathways through the community, indicating the routes that nutrients, and pollutants, might follow. These are data-intensive models, requiring considerable work to establish all the connections and their magnitude. The complications include:

- seasonal or transitory species visiting to feed before moving on, and taking energy and nutrients with them;
- communities that cycle between different states, creating local patches where different assemblages (with different dominant species) persist for relatively short periods. This creates problems in sampling and representing the variation in the web model;
- change over time and in response to abiotic changes, as in a succession and the development of ecosystem processes. Many food web models assume a community undergoing no successional development.

The behaviour of the ecosystem, and the response of its food web to a disturbance, is a potential indicator of community integration. Communities that change readily, and suffer major shifts in their energetics, suggest some measure of instability. Does this indicate a high or low degree of integration: a community that has co-evolved for some time, or one in the early stages of a succession?

➔ *Section 6.1, Section 6.4, Community integration and functional redundancy*

Box 7.3 Key technique

Measuring energy transfers and food webs

Partitioning gross and net primary production is inevitably the first step in studying the energetics of an ecosystem. Considerable effort is needed to account for all energy fixed by the system, given that the large and obvious primary producers will not be the only sources utilized by consumers.

The methods for quantifying primary production are usually based on gravimetric measures of biomass and standing crop, but the details depend on the nature of the ecosystem and its autotrophs. Measuring incident radiation, integrated over the year, allows derivation of photosynthetic efficiencies. Gaseous exchange—rates of carbon dioxide uptake or oxygen production, with and without illumination—allow calculation of photosynthesis and respiration rates. Similarly, gravimetric techniques and respirometry can be used to measure secondary production and its efficiency by particular species, or in some cases (such as soils), for entire communities.

The net primary production passing to the grazing and decomposer food chains can be derived from stable-isotope analyses or direct observation methods (such as exclusion studies or feeding trials), and similar techniques can be used to measure transfer efficiencies along food chains.

However, these require a full description of the consumer pathways (including alternative food sources) and a partitioning of species between trophic levels. Such compartmentalization has to properly weight feeding relations, distinguishing between the infrequent and the dominant energy pathways.

A food web establishes what eats what. Field work observing feeding relations, stomach contents, or faecal residues is feasible for large consumers. Although it is the most direct method it will rarely be definitive. For smaller animals, radioactive tracers (stable isotopes of key nutrients) can help to confirm and quantify feeding relations.

Immunological techniques (of gut contents or residues) can also be confirmatory, but increasingly DNA or protein analysis provides a faster alternative. With these methods repeated measures are required to indicate the strength of the connection.

The counts from a radioactive tracer can measure the speed with which a nutrient passes through a web and from one trophic level to another. With radiolabelled carbon we can also determine rates of loss through respiration. The stable isotope of hydrogen (deuterium) has been used to establish both feeding links and their strength in aquatic ecosystems: levels in the diet, rather than the water, are represented in a consumer's tissues.

continued

The efficiency of energy transfer along a food chain requires detailed study of individual organisms. The energy in tissues or faeces can be measured in a bomb calorimeter, burning the biomass in an excess of oxygen and recording the heat produced. Respirometry is used to find basal metabolic rates and to estimate the efficiency of production in both primary and secondary producers.

➔ *Box 2.1, Energy budgets*

Looking for extra marks?

The properties of food webs and the properties of species

We might expect the complexity of food webs and the length of food chains to depend on the species assemblage, their assimilation efficiencies and physiologies.

Consider which food chains are likely to be the longest—perhaps those dominated by poikilotherms, with their higher assimilation efficiency—such as those comprising insects? Should food webs with a high turnover rate support more mammals with their low assimilation efficiency, such as the African savanna? Does the complexity of the web decrease when nutrients become locked away, such as in cold and waterlogged soils?

Ensure you know how the food web develops in a succession you have studied in detail. Do the energetic efficiencies change as the food web develops?

7.3 STABILITY AND FOOD WEB STRUCTURE

A disturbance to a food web might disrupt:

- energy transfers,
- nutrient transfers,
- species associations.

Complexity was thought to impart high stability in diverse communities, since a species-rich community offered alternative pathways for energy and nutrients to pass through a food web.

This complexity is the product of:

- the number of species,
- the number of connections between them,
- the frequency or strength of the connections.

The **connectance** of a community is the proportion of all possible routes through the system that are used. High connectivity means energy or nutrients pass down most possible routes.

It was suggested that the loss or addition of species in highly connected communities would have little effect on these transfers since alternative pathways would dampen any disruption to the larger system. This was supported by several observations:

- pest outbreaks are more prevalent in species-poor agricultural communities;
- populations are more variable in the less diverse ecosystems of temperate regions, compared to the tropics;
- species-rich communities appear less easily invaded by new species compared to others.

An alternative view is that highly integrated communities might show less resilience and elasticity (Figure 7.3): high connectance could mean a local disruption was propagated across the food web. Rather like a wave, it would be transmitted to most corners of the web and back again.

Mathematical models that vary the number of species, and the strength of connections between them, suggest that stability does not follow from high species richness. Compared to more loosely configured early successional communities, the highly connected web of a late succession appears less stable, or at least shows low elasticity.

From both models and comparisons of real food webs, it seems that:

- a food web with increasing species richness will only remain stable if the number of connections, or their average strength, falls;
- most complex food webs appear to have only a small number of strong links;
- a reduced connectivity dampens the impact of a disturbance so it is not propagated through the community.

Strong interactions exist between species with similar functional roles—*within* **guilds**—between species likely to be competing or responding to similar selective pressures. The links *between* guilds are likely to be relatively weak.

➔ *Section 6.4, Species–area relationships*

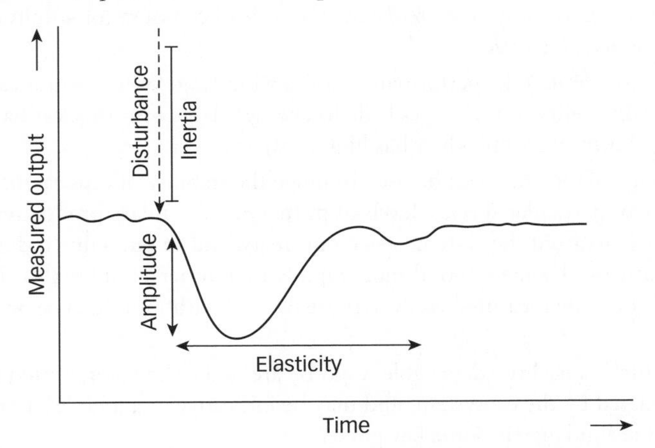

Figure 7.3 Measures of stability. The capacity of the system to remain unchanged with different scales of disturbance is its inertial stability (or resistance). The size of the deflection (or amplitude) in response to a particular scale of disturbance is its resilience; the time taken to recover its former level is its elasticity. A system not easily deflected and which recovers quickly from some disturbance is likely to have high functional redundancy.

Stability and food web structure

A guild provides functional redundancy for the role it performs, so that disruption within the guild is not transmitted to the larger food web. This damping effect is greater when the trophic distance between guilds is large. This property has been described for nitrogen transfer in soil communities and among the guilds of coral reef fish.

Box 7.4 Key technique

Concepts of stability

Homeostasis attempts to maintain the internal environment of cells and individuals within certain limits, regulated by negative feedback, through processes evolved to maintain function at minimal cost to the organism. Ecosystems are also regulated by negative feedback, but their constancy is a product of co-evolution, where each species seeks to maintain its reproductive fitness, evolving in an evolving biotic environment.

➔ *Section 5.1, Species interactions*

Just like a cell, we can measure the stability of populations, communities, and ecosystems. **Stability** is a property of systems that change little following a disturbance, or which return quickly to their previous condition. This comprises (Figure 7.3):

- **inertial stability**: the capacity to resist deflection when disturbed;
- **adjustment stability** (or elasticity): the speed of return to its original condition after being disturbed.

We can also measure a system's *resilience*: the scale of response (or amplitude) to a disturbance of a given size.

A cell able to maintain its performance over a wide range of temperatures shows high inertial stability; if it changes little following a large heat shock it has high resilience; if it recovers quickly it has high elasticity.

A wide range of measures can be used to judge the stability of ecosystems: the abundance of particular species, levels of primary or secondary productivity, the loss of nutrients from the system, species diversity, and so on. Some indicators are more sensitive and some respond more rapidly than others, but we also choose metrics that can be measured easily, repeatedly, and without influence on the system.

Stresses which cause little detectable response are said to be incorporated or accommodated by the ecosystem, and may be indicative of a high level of functional redundancy in some key guilds.

➔ *Section 6.4, Incorporation and succession, Functional redundancy*

7.4 THE BIOGEOCHEMICAL CYCLES

The speed at which an ecosystem is working is indicated by the transit times of key nutrients. An atom of calcium will pass through a northern coniferous forest in 43 years and a tropical rainforest in just over 10 years. The major difference is the abiotic forcing functions (temperature, energy, and water availability) that govern the rates of biological processes.

Compartment models enable us to compare these **flux rates** between ecosystems (Figure 7.4). A passage through the **biosphere** diverts, slows, and—in some cases—accelerates nutrient flux rates.

- Rates differ between elements according to their chemistry and physical properties.
- Their atmospheric residence time depends on the particle size and solubility of their common compounds.
- Energy captured in primary production powers the cycling of significant quantities of the key nutrients.

This is demonstrated by nitrogen. Life allows nitrogen to cycle faster, and fixed nitrogen supports carbon capture by the biota.

- High rates of photosynthesis require high levels of leaf nitrogen (for the photosynthetic enzymes and chloroplasts). Symbiotic nitrogen-fixing bacteria require a carbon and energy subsidy from the host plant to make nitrogen fixation cost-effective.

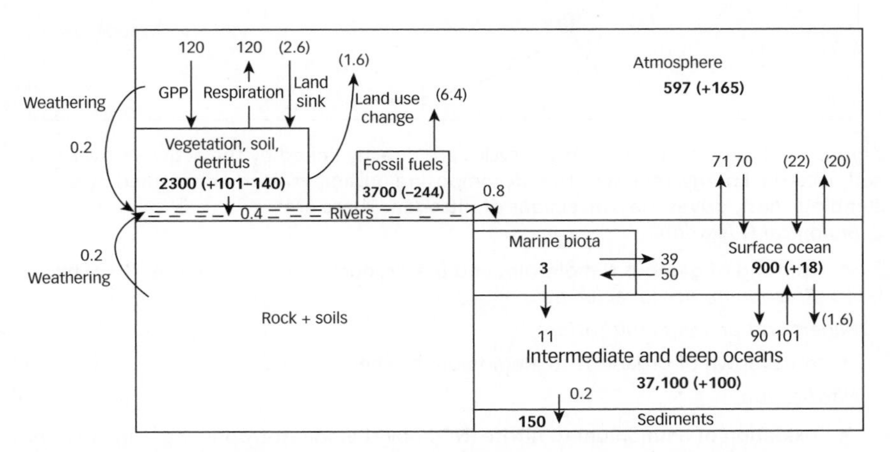

Figure 7.4 A model of the global carbon cycle, using data from the UNEP IPCC assessment (2007) for the year end 1994 (gigatonnes (Gt or 10^{12} kg) are shown in each compartment; Gt/year for flux rates).
Numbers against an arrow show the pre-industrial flux rates (those in parentheses give rates attributable to industrial activity).
Numbers standing alone represent the mass in that compartment (numbers in parentheses give the mass added or lost due to human activity since 1750).
GPP, gross primary production.

- The decomposer food chain provides the energy to oxidize and reduce compounds of nitrogen added to the soil. This supports all protein synthesis in the community, including that needed for photosynthesis. The carbon and nitrogen cycles depend on each other and are the basis of symbiotic associations fundamental to most ecosystems (Figure 7.5).

Generally, nutrients accumulate in ecosystems as a succession proceeds, and as organic matter builds up. The amount available to its biota is the nutrient capital of an ecosystem (Table 7.1).

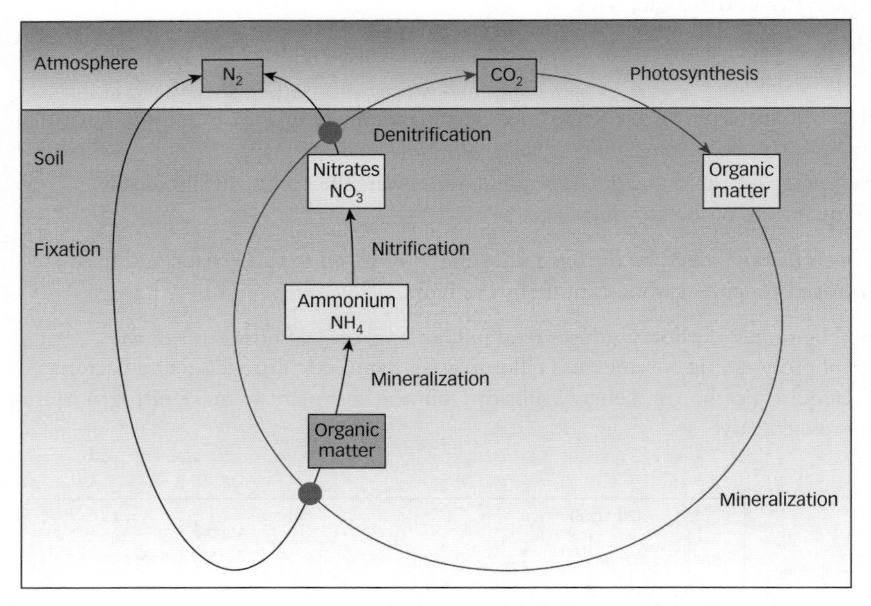

Figure 7.5 The carbon and nitrogen cycles are directly linked by the activity of various soil bacteria. Energy, released from decomposing organic matter, or provided by a symbiotic host, drives the conversions of nitrogen. These are:

Biological N fixation:

- splitting of gaseous N molecules and their reduction to ammonium (NH_4^+) by *Rhizobium, Azotobacter,* etc.

Mineralization (ammonification):

- breakdown of organic N to ammonium by chemoheterotrophs.

Nitrification:

- oxidation of ammonium to nitrite (NO_2^-) by chemoautotrophs, e.g. *Nitrosomonas;*
- oxidation of nitrite to nitrate (NO_3^-) by chemoautotrophs, e.g. *Nitrobacter.*

Denitrification:

- reduction of nitrate to gaseous N by heterotrophic bacteria under anaerobic conditions, to oxidize organic matter, e.g. *Bacillus, Pseudomonas.*

Nitrate, assimilated by higher plants, bacteria, fungi, and blue-green algae, is reduced to ammonium in the synthesis of amino acids.

The filled circles indicate two key interactions of these cycles: at N fixation and denitrification, when N enters and leaves the soil.

	Input into biota	*Long-term sink*	*Output to atmosphere*
Entering the biota from the atmosphere			
Carbon	Principally as CO_2 via photosynthesis	Sedimentary rocks (forests, soils, and sediments in the short term but become long term when fossilized)	Respiration, fermentation, fossil fuel combustion, industrial processes, vulcanicity, agriculture
Nitrogen	Into solution through electrical storms; bacterial, fungal, and algal fixation direct from atmosphere	Sediments, biota, atmosphere	Bacterial denitrification, fossil fuel combustion, industrial processes, artificial fertilizers, animal wastes especially from agriculture
Entering the biota as deposits arriving principally from the atmosphere			
Sulfur	Wet and dry deposition from atmosphere assimilated by bacteria, fungi, phytoplankton, and higher plants	Sediments	Vulcanicity, fossil fuel combustion, mining and industrial processes, dimethyl sulfide release from phytoplankton, bacterial H_2S release
Entering the biota from sediments or rocks			
Phosphorus	Chemical weathering of rocks, fungal release from organic matter, algal and phytoplankton absorption from water	Sedimentary rocks, soils, and sediments	Decomposition, fishing, and artificial fertilizers
Potassium	From soil water, weathering of rocks	Sediments, organic matter	Decomposition, artificial fertilizers, agricultural animal wastes

Table 7.1 The biogeochemical cycles of the major nutrients.

- The most important biological interactions for these cycles involve the microbial community, especially within the decomposer food chain.
- Community development, nutrient storage, and flux rates are closely linked, drawing on the variety of metabolisms of the bacteria and fungi.
- The interaction of these transfers, especially among saprotrophs, determines the availability of limiting nutrients and the productivity of the larger community.

➔ *Section 6.2, Nitrogen and phosphorus availability in soils during plant successions*

Elaborate food webs are also found in ecosystems receiving no radiant energy. These are built on the primary production of chemoautoptrophic bacteria and can represent a considerable biomass. The most spectacular are found around the hydrothermal vents in deep, dark oceanic trenches. These 'black smokers' are a source of reduced sulfur and many other compounds, and the chemical energy on which the whole community depends.

Each of the major nutrients has a global cycle which may include geological and biological sinks.

The biogeochemical cycles

- The concentration and residence times for some elements in the atmosphere determine the abiotic conditions in the biosphere and the metabolism of its living components.
- These cycles have changed radically in geological time. For example, the fossil fuels we exploit today were deposited in the warm and wet Carboniferous period, during which vast amounts of carbon were removed from the air by photosynthesis.
- The productivity of the planet is set by nutrient availability and flux rates in the biosphere.

Revision tip

Ensure you know the essential features of the main nutrient cycles, and their interactions with the biota. You will also need to know the hydrological cycle and the cycle for oxygen.

Nutrient cycling in climax communities

The mineral balance of climax communities was thought to be close to equilibrium, providing relatively constant nutrient reserves. Based mainly on studies of terrestrial plant communities, the suggestion was of:

- low loss of minerals, especially the key plant nutrients (N, P, K),
- high rates of internal cycling of these nutrients,
- high rates of capture of additional nutrients.

Each of these features was thought to follow from the highly competitive plant community that developed as the nutrient supply became limiting.

➔ *Section 6.2, Development of ecosystem system function*

This view has been challenged. Some argue that close control over nutrients requires high primary productivity, since:

- high nutrient capture occurs when biomass is increasing rapidly and energy is available;
- low levels of primary production allow fewer nutrients and less energy to be added to the soil as organic matter, reducing availability to growing plants.

In any community, the internal cycling of nutrients depends on their release from different storage compartments. Decomposition, the release of energy and nutrients from detritus, is critical to the productivity of terrestrial ecosystems (Figure 7.4), where rates are set principally by ambient temperature and water availability.

➔ *Section 8.3, Abiotic characteristics of the major biomes*

Looking for extra marks?

Useful comparisons can again be made between temperate and tropical systems: consider comparing wetland systems such a cypress or mangrove swamp with the salt marshes of temperate latitudes. Here differences in incident radiation, temperature, tidal range, and salinity are the prime external variables governing rates of decomposition, primary production, and nutrient flux in comparisons between latitudes.

Low nutrient availability and reduced primary production can create the mosaic of habitat patches seen in many terrestrial plant communities. Losses through periodic disturbances may reduce nutrient availability, creating a subclimax. A patch then has to re-establish the processes and nutrient capital associated with a climax community.

- Different patches represent local balances between primary production and nutrient availability (and storage).
- The community may cycle between these states, prompted by disturbance and followed by autogenic interactions.
- In other systems, small differences in abiotic conditions can create this patchiness. In temperate freshwater streams, primary production is concentrated in particular areas (**riffle**), although these play a minor role in the nutrient cycling of the larger system.
- Streams are dominated by their decomposer food chain, but are being continually re-moulded with changes in water flows. Organic matter is readily lost with large discharges.
- The low nutrient capital of streams reflects their seasonal input of allocthonous organic matter. The invertebrate community comprises species with a life history reflecting this seasonality.

➔ *Section 6.2, Mosaic communities and patch dynamics*

7.5 BIOGEOCHEMISTRY AND THE GLOBAL ENVIRONMENT

Since the arrival of the photoautotrophs over 2.5 billion years ago, the composition of the Earth's atmosphere has been affected by the biological activity in its biosphere.

- The advent of photosynthesis by bacteria began the enrichment of oxygen in the atmosphere, reaching concentrations which could support faster and more efficient energy release—aerobic respiration—ultimately allowing some organisms to chase their food.
- The evolution of higher plants and animals became possible as multicellular organisms could exploit free oxygen for their catabolism.

Make the connection

Given the origins of oxygen in the atmosphere, note the range of catabolic metabolisms, and their phylogeny, described in Section 7.2. Consider especially the types of respiration among the autotrophs, and the early history of life on Earth.

The effects of nutrient cycling on the global abiotic environment are most evident with the carbon cycle and the greenhouse effect: the link between the mean atmospheric temperature and the concentration of carbon and water vapour in the atmosphere.

- Atmospheric gases, particulates, and aerosols determine which wavelengths of radiant energy reach the planet's surface.
- The longer, infrared wavelengths reflected back by the surface are absorbed by the atmosphere, and most especially its water vapour and carbon-rich gases.
- Re-radiation of this energy into space occurs in the upper atmosphere where it is cold. At these lower temperatures, the heat loss is slow. This is the greenhouse effect: the temperature rises because energy is lost more slowly than it is received.
- Without this effect the mean atmospheric temperature would be −17°C, rather than its current average of 15°C.
- Carbon dioxide is not as effective an absorber of these wavelengths as some other gases, but its relatively high concentration means it accounts for about half the greenhouse effect.
- Because carbon dioxide is accumulating in the atmosphere, its residence time is increasing. This makes it the most important greenhouse gas.
- The carbon content of the atmosphere is primarily regulated by the biosphere: fixed by photosynthesis and released by respiration and decomposition. Temperature fluctuations are 'amplified' because of the positive feedback of temperature on biological activity.
- If nothing else changes, a warm climate increases carbon release by respiration and decomposition; it also increases the flux of water through the atmosphere, itself an important greenhouse gas and a means of 'holding' energy in the atmosphere.
- Carbon is removed from the atmosphere through non-biological absorption by the oceans. However, the most important loss is through photosynthesis (Figure 7.4).
- Terrestrial ecosystems represent more than two-thirds of the carbon fixed in primary production each year, and two-thirds of this is fixed by forests. Tropical forests capture almost half of all atmospheric carbon fixed on land.
- Carbon fixed in living tissues may not be released, but instead becomes incorporated into sediments.

- One important mediator is the marine phytoplankton, whose photosynthesis and productivity rise with temperature. They remove significant amounts of carbon rapidly from the system by settling out in the sediments. The same happens with the death of marine invertebrates that use carbon dioxide and calcium to build their shells and exoskeletons.

The burning of fossil fuels increases the concentration of carbon dioxide in the atmosphere and deforestation reduces the rate at which it can be fixed by photosynthesis. The residence time of carbon dioxide in the atmosphere rises as a result and so do mean atmospheric temperatures.

Make the connection

Genetic change through global change

We can link the rise in atmospheric carbon with changes at the molecular level: in the nucleotides comprising the genomes of polar bear and grizzly bear populations in North America.

With higher atmospheric temperatures and the retreat of the Arctic ice sheet, the two bears increasingly overlap in their ecological space. Possibly sympatric species, their gene pools today exchange information more frequently.

The incidence of grolar bears suggests the small genetic differences between them are melting away as polar bears have to feed further south.

Section 1.2, Ecological niche

Check your understanding

Examination-type questions

1. Describe how the global cycling of carbon and nitrogen are linked by the biota of a terrestrial plant community.
2. During one growing season 114 720 kJ/m² of radiant sunlight fell on a forest in a watershed at the Hubbard Brook Experimental Station in New Hampshire, USA. Of this 2524 kJ was fixed as gross primary production and 1377 kJ was respired by the plants.
   a. What was the gross primary production efficiency of the plants?
   b. What the net primary production efficiency of the plants?

   The standing crop of the living plants represented 17 069 kJ/m².

   c. What proportion of the living standing crop does the net primary production of that growing season represent?
   d. What is the turnover time for the energy in this standing crop?

   Herbivores consumed 10 kJ/m² during the growing season. Some 830 kJ/m² was available to detritivores.

e. What proportion of the incident radiation was consumed by the primary consumers on the grazing food chain?

f. What proportion of the incident radiation was available to the primary consumers on the decomposer food chain?

Litter fall during the growing season added 726 kJ/m^2 to the soil surface. The litter had a standing crop of 8203 kJ/m^2. The organic matter in the upper soil (beneath the litter layer) represented 21 061 kJ/m^2.

g. What is the turnover time for the standing crop represented by the litter?

h. What is the total amount of energy available to all consumers in this ecosystem (in each square metre)?

i. What proportion of this energy is available to each food chain?

(These data are from Gosz, J. *et al. Scientific American* **238**(3):92–102, 1978.)

You'll find answers to these questions—plus additional exercises and multiple-choice questions—in the Online Resource Centre accompanying this revision guide. Go to http://www.oxfordtextbooks.co.uk/orc/thrive or scan this image:

8 The ecology of the earth

Key concepts

- The position of the Earth, its size, and orbit are part of the explanation for the development of life on Earth. The search for extraterrestrial life attempts to find extra-solar planets that meet similar criteria.
- The ecology of the Earth developed as its biota evolved, with changes to the atmosphere, the cycling of nutrients, and the formation of soils and landscapes.
- The distribution and phylogeny of many organisms can be related to change over geological time, such as continental drift.
- At least five mass extinctions are recorded in the fossil record, although species richness has increased progressively through geological time. The Earth is now losing species at the fastest rate in its history.
- The different ecosystems across the planet reflect major abiotic gradients, most especially temperature. On land, climate and the availability of water govern the distribution of the regional plant communities or biomes.
- Temperate and sub-Arctic biomes are relatively young ecosystems, formed since the end of the Pleistocene glaciations.
- A range of explanations are offered for the gradient in species richness between the poles and the tropics, in both terrestrial and aquatic ecosystems. The most obvious is the higher primary productivity in the lower latitudes.

- Many ecosystems are stratified by local abiotic gradients, zones distinguished by their ecological communities and processes.
- Soils are structured by gradients operating at regional scales, but also reflect local topography, geology, and history. Soil fertility tends to decline from the temperate biomes to the tropics.

Assumed knowledge

The geological time scale. Continental drift. The climatic patterns of the Earth. Basic astronomy and the origin of the tides.

A conceptual map for this chapter is available on the companion website; go to http://www.oxfordtextbooks.co.uk/orc/thrive/ or scan this image:

8.1 ECOLOGIES OF SCALE

The Earth is the only body in the solar system known to support life and thus the only indication we have of the necessary conditions for life to evolve. Several of these conditions follow from the planet's position on the gradient of radiation with distance from the sun.

Gradients in space

Exobiology

Claims have been made for several 'Earth-like' planets orbiting other stars, even though they cannot be observed directly. Astronomers estimate the planet's distance from its star by the orbital period, and the composition of any atmosphere by the wavelengths of light absorbed by the planet.

- Not only will it be hotter closer to the star, the force of the solar wind will be greater: this stream of charged particles can strip away an atmosphere. A planet has to have sufficient mass to retain an atmosphere. For both these reasons tiny Mercury has none.
- Atmospheric composition is also critical: no life is thought to exist on Venus where the dominant gas is carbon dioxide. Its high temperatures and pressures are beyond the range of any biological processes described on Earth.
- Since the advent of photosynthesis, the Earth's atmosphere has protected organisms from harmful radiation. The ozone layer in the stratosphere absorbs incoming ultraviolet radiation that can damage structural proteins and DNA.
- Liquid water is assumed to be a prerequisite for life, needed to recreate the biochemistries we see here. It also suggests the extra-solar planet has a temperature range similar to the Earth's. High atmospheric concentrations of water and abundant carbon dioxide help buffer mean surface temperatures.

- The hydrological cycle is crucial in redistributing energy through the Earth's weather systems. This helps reduce regional differences in temperature on both a daily and a seasonal basis.
- Received solar radiation varies with distance from the sun, the rate of spin of the planet (its daylength), and with the seasons, due to the tilt in the axis of spin. Unequal warming on Earth also arises from the unequal distribution of landmasses.
- Variations in the tilt, shape, and precession of the orbit all create long-term cycles. These **Milankovitch cycles** are correlated with major shifts in the climate and biomes of the Earth.
- Over a shorter period, the gravitational effects of the Moon create the tidal range in sea levels found at higher latitudes, although this too has changed during the evolution of the Earth.

Such astronomical effects can now be measured for planets that are inferred to exist around other stars. Even before we have seen the planet we can begin to describe some of the key abiotic factors that would feature were any ecosystems to develop.

Evolution of the Earth

Ecologists need a perspective in time, as well as space, to understand how life has evolved and ecosystems have developed. We need to know the temporal and spatial scales over which key processes have an impact (Table 8.1).

- Unicellular life appeared on Earth within a billion years of its formation.
- With the evolution of photosynthesis, around 2.5 billion years ago, the reducing atmosphere changed to one with a high proportion of oxygen.
- As primary productivity fixed more energy, nutrients began to cycle faster.
- Multicellular life appeared in the oceans 550 million years ago. Twenty million years later, the Cambrian 'explosion' records a rapid increase in species diversity and the appearance of the major animal phyla extant today.

	Large spatial scale	**Small spatial scale**
Large temporal scale	Continental drift and the zoogeographical regions	Evolution on islands
Small temporal scale	Tides and diurnal changes in water flow and salinity	Pulses of nutrients or pollutants and short-lived changes in ecosystems, e.g. inputs in freshwater streams

Table 8.1 Examples of large- and small-scale processes that help explain the ecology of life on Earth.

Large temporal scales are associated with phylogenetic differences and evolutionary change; factors operating over large spatial scales determine the nature of regional ecosystems.

Small-scale processes may determine the local organization and performance of an ecosystem, and perhaps its longevity.

Looking for extra marks?

The Cambrian explosion

You should be aware of the significance of the Cambrian explosion and the possible explanations for such a rapid diversification over a relatively short period of time.

The development of regulatory genes controlling the spatial organization of multicellar animals (the homeotic genes) allowed patterns to be repeated down their body length, an innovation that gave rise to the major animal phyla. Thus the appearance of segmented bodies—in higher invertebrates and subsequently the vertebrates—led to a fundamental shift in the variety of life.

Thus the nature of the planet's ecosystems and the diversity of their communities changed following a small mutation, but one with immense adaptive significance.

Three trends are apparent from the fossil record since the Cambrian:

- an increase in the complexity of multicellular organisms;
- an increase in the abundance and species richness of multicellular organisms;
- an increase in the range of ecosystems and the abiotic gradients they occupy.

Together these also indicate an increased role for ecosystems in the storage and cycling of elements.

These changes are a consequence of evolution because:

- growth and reproduction require energy and other resources;
- populations evolve and adapt as habitats change with time;
- species adapt and exploit new abiotic environments;
- with co-evolution, the biotic environment changes as species change;
- species went extinct as others appeared.

The mass extinctions in the fossil record

In different parts of the world, sedimentary deposits record the same discrete events and transitions over geological time. The fossil record shows:

- five or six mass extinction events with the loss of a significant proportion of multicellular species in a short period. Each is attributed to a catastrophic change in the global environment;
- up to 95% of oceanic species disappear during a mass extinction, but numbers have always recovered quickly;
- however, most extinctions occur outside these events: around 90% of losses are part of a background rate of species turnover.

The fossil record also shows that the number of species has consistently risen as the Earth has aged.

- Currently, the oceans have a variety of animals about twice the average for much of the fossil record.
- Most estimates range between 5 and 10 million species extant on the Earth, with the majority uncounted and unclassified.
- Today the Earth has the highest species diversity in its history, but is losing species faster than it has ever done before: we are living through its greatest extinction event.

Make the connection

Besides the difficulties of finding them, consider some of the taxonomic problems in counting all species alive today, and their significance for our estimates of current diversity and extinction rates.

Note especially the value of an internationally accepted protocol for their naming and classification…and for a system of type specimens.

 *Section 1.1, Taxonomy and phylogeny*

The most likely cause of the current mass extinction is the speed with which human activity is changing the environment. To take one example:

- Corals and shelled marine invertebrates have one of the most complete fossil records. They are ancient groups whose phylogeny and diversity are easily followed since the Cambrian explosion.
- Reef communities are complex and readily disturbed. Coral species are often highly localized and may disappear when its single population goes extinct.
- However, as one dominant coral species is lost, another quickly replaces it. These communities have shown rapid recovery after each extinction event, restoring their species richness within 5–10 million years.
- Similarly, there have been few extinctions among tropical reef biota over the last 2 million years, despite major fluctuations in sea level. Instead, these communities have tracked the shift in the climatic regions, moving with them.
- With the fall in sea temperatures during the ice ages the extinction rate of Caribbean molluscs increased markedly, but this was more than offset by an increase in speciation rates.
- This high elasticity is probably indicative of high functional redundancy associated with the species guilds described for these ecosystems.
- Similarly, the devastation of Caribbean manatee, turtle, jewfish, and conch populations by human predation had little effect on its reef ecosystems over the last 500 years. Other species, primarily invertebrates, increased their abundance to exploit the resource opportunities.
- More recently, the elastic limit of reef ecosystems has been reached. Increased sedimentation and pollution have contributed to widespread coral death in the Caribbean. These ecosystems now show signs of permanent change.

Ecologies of scale

Coral bleaching and the collapse of reef ecosystems is now a global phenomenon, associated with water temperatures persistently above their long-term average.

➔ *Section 7.3, Elasticity in ecosystems*

> **Looking for extra marks?**
>
> The protection of habitats needs to be informed by the detail of the ecology of the constituent species, their interactions with each other, and their abiotic environment.
> Answers need this detail to justify conservation measures for any threatened ecosystems you have studied. In this example, you would need to explain why high water temperatures should cause bleaching in coral reefs.

Large-scale spatial change and evolution

Not only has the atmosphere of the planet changed with time, so has its geography. Continents drifted across the globe, taking their plants and animals with them. These movements over the last 200 million years can be followed in the distribution of terrestrial plants and animals which evolved during this time.

This period encompasses the rise of the mammals and the world can be partitioned into six zoogeographical realms based on their mammalian fauna (Table 8.2). These were first described by Alfred Russel Wallace though, ironically, they posed a significant problem for evolution theory:

- closely related taxa, sharing a recent ancestor, should share the same or adjacent landmasses, yet some were separated by half the globe;
- living marsupials are found only in the Americas and Australasia. Given their unique reproductive anatomy, and the unlikelihood of convergent evolution, how could pouched mammals be so widely separated?

Alfred Wegener suggested this (and other anomalies) indicated that continents were once connected. Rather than a series of convenient (but implausible) land bridges, he proposed that continents moved.

- His theory of continental drift was finally accepted in the 1960s with the discovery of the mechanism powering this movement. Plate tectonics describes how the heat of radioactivity in the Earth's core creates convection currents in the mantle rocks. These push the crustal plates around its surface.
- The first mammals appeared as the single large plate of Pangaea began to break up. This divided into a northerly continent (Laurasia: North America, Europe, and most of Asia) and a southern landmass (Gondwanaland: the southern continents including Antarctica).
- Fossils from China suggest the placental mammals (Eutheria) and the marsupials (Metatheria) diverged in Laurasia, around 160 million years ago.
- Today the more ancient marsupial taxa are found together in the Americas, (principally the opossums of North and South America), as are their fossils.

- The youngest and most diverse marsupial groups are found in Australia and New Guinea, which together split from Antarctica 65 million years ago.
- The marsupials reached Australasia via Antarctica, (where fossil specimens have also been collected).
- The timing of the break-up of Gondwanaland corresponds with the distribution of extinct and extant marsupials, and the distribution of fossils of taxa of different ages. It also explains the absence of marsupials from continents isolated earlier.

➔ *Section 2.3, Evolution and phylogeny*

Holarctic (broadly equivalent to Laurasia)	
Palaearctic	Certain hominids
Nearctic	Certain marsupials (*no primates*)
Tropical (broadly equivalent to Gondwanaland)	
African	Hyraxes, hippos, prosimians (galagos, lemurs, etc), Catarrhini (Old World monkeys and apes), certain hominoids (*no edentates or Platyrrhinni*)
Neotropical	Edentates (armadillo, sloths, anteaters), certain marsupials, Platyrrhini (New World monkeys) (*no bovids, prosimians, or Catarrhini*)
Indomalayan	Certain monotremes and marsupials, certain Prosimians (lorises, tarsiers, etc.) and Catarrhini
Australian	Certain monotremes and marsupials (*no placentals except rodents*)

Table 8.2 The distinctive mammalian fauna of the six zoogeographical regions.

Reflecting their almost continuous contact in the last 200 million years, the Palaearctic (Europe and Northern Asia) and the Nearctic (North America) share most groups, and together are called the Holarctic.

Because they separated around 70 million years ago, Africa and the Neotropical (South America) have very distinct mammalian faunas, especially in their primates. The Indomalayan region shares groups with its neighbours. The absence of placentals and the dominance of the marsupials gives Australia the most unique mammalian fauna.

Much of the mammalian fauna of the ancient northern continent is common between its two realms because there have been no significant barriers between them. They are collectively referred to as the Holarctic (Table 8.2).

- Gondwanaland split from Laurasia 140 million years ago. Mammal groups which evolved afterwards show restricted distributions.
- Laurasia itself divided 20 million years ago. Consequently there are no endemic primates (apes, monkeys, and their relatives) in the Nearctic.
- Human evolution and dispersal reflects the glacial advances of the last 1.6 million years, during the Quaternary period. The migration of our subspecies out of Africa began around 100000 years ago, before the last glacial advance of the Pleistocene epoch.

Recent change and ecology

The current distribution of terrestrial plant communities reflects the Earth's more recent past and the prevailing climate of the Holocene. Broadly following lines of

latitude, major plant assemblages are organized into **biomes**, species sharing adaptations to the regional climate (Figure 8.1).

A Classification of the Major Ecosystems

Terrestrial Ecosystems

Forest

Warm, wet → ***Cold, wet***

TROPICAL RAIN FOREST | TEMPERATE DECIDUOUS | COASTAL CONIFEROUS | TAIGA

Scrubland

Warm, seasonally dry → ***Cold, seasonally wet***

CHAPARRAL, MAQUIS | FYNBOS | HEATH AND MOOR | TUNDRA

Grassland

Warm, seasonally wet → ***Cold. seasonally wet***

SAVANNA | PRAIRIE, PAMPAS, STEPPE | TUNDRA

Desert

Hot → ***Cold***

SAGEBRUSH | SCRUB

Aquatic Ecosystems

FRESHWATER

Lentic

Large → ***Small***

INLAND SEAS | LAKES | PONDS FEN, INLAND SWAMP

Lotic

Large → ***Small***

RIVERS | STREAMS | INLAND MARSH

MARINE

Littoral

High erosion → ***High Deposition***

ROCK PLATFORM BOULDER FIELD | SHINGLE BANKS | BEACHES | SALTMARSH, SWAMP

Sub-littoral

Rock | ***sediment*** | ***Biological construction***

KELP FORESTS | SEA GRASS FIELDS OR SOFT-BODIED ANIMALS | REEFS

Oceanic

Vertical gradients

SURFACE (NEUSTIC) → BENTHIC | ABYSSAL AND HADAL DEPTHS

Horizontal gradients

CONTINENTAL SHELF (NERITIC) → OPEN OCEAN

Figure 8.1 A single gradient is used to define each ecosystem because of its significance for other abiotic factors; for example, in littoral ecosystems high exposure determines rates of erosion and deposition, average particle size, and the loss or build-up of nutrients.

This simple classification ignores transitional communities where one type develops into another and only includes the most persistent ecosystems.

- Within a biome most plants have a characteristic growth form adapted to the prevailing abiotic conditions.
- We can distinguish four major terrestrial biomes by their regimes of precipitation and temperature:
  - *forests* dominate where there is sufficient moisture;
  - *grassland* or low scrub where it is drier;
  - *deserts* are found where evaporation exceeds rainfall;
  - *tundra* in the sub-Arctic where the ground is frozen for much of the year.
- Local conditions may confound the simple latitudinal pattern of the biomes. Proximity to an ocean or a high mountain range can create local climates that support unique, and isolated, plant communities.

Revision tip

Know your biomes

You will need to know the dominant plants in the major terrestrial biomes, and be able to relate their adaptations to the regimes of temperature and moisture; similarly the autecology of the characteristic animals.

Note that there are equivalent regions in the oceans, defined by their latitude, depth, and position in the major currents.

8.2 GLOBAL GRADIENTS OF SPECIES RICHNESS

The diversity of many groups increases from the poles to the equator, so the greatest number of species per unit area is found between the tropics.

- For example, a range of benthic fish and invertebrate groups (especially molluscs and crustaceans) show this gradient in the North Atlantic. This pattern is less distinct in the South Atlantic and Indo-Pacific oceans.
- Tropical estuaries and deep sea communities have higher species richness than their counterparts in the higher latitudes.
- The contrast is more obvious in terrestrial biomes: 70% of all multicellular species are found on 6% of the Earth's land surface, in its tropical forests.
- Some biologists have also proposed latitudinal trends of anatomy or niche space within groups of species, reflecting gradients of temperature and productivity (Box 8.1).

Box 8.1 Make the connection

Global patterns in adaptation

Global abiotic gradients may be indicated by adaptations within a group of organisms.

Bergman's rule (or the size rule) suggests colder regions have populations with larger body sizes than those in warmer climates. Bodies with a smaller surface-to-volume ratio lose heat less rapidly. The rule applies to many homoiothermic

continued

animals, and seems to work best for birds and mammals, or when comparing within-species variation. It does not apply so readily to poikilotherms.

The result is a **cline**—a character gradient—of increasing size with latitude and sometimes altitude. A good example is the wolves of North America and Europe. However, some small mammals, such as shrews, show the opposite pattern, possibly due to the food limitations on insectivores in highly seasonal habitats.

Allen's rule suggests that extremities—limbs, ears, tails, etc.—are shorter in cooler areas, again to reduce heat loss. However, this seems to apply only to a small number of homoiotherms.

Rapoport's rule says that plants and animals of lower latitudes are confined to smaller latitudinal and altitudinal ranges. This suggests the greater diversity in the tropics is a result of narrower ecological ranges and greater species packing within an area. For some groups this appears to hold (many vertebrates and trees) but for others the evidence is not supportive. It could also reflect the wide range of tolerances needed for life outside the tropics or the reduced competition and niche specialization at these higher latitudes.

Is there a prime abiotic factor to explain these patterns in both marine and terrestrial communities? The principle theories are summarized in Table 8.3, but it is perhaps a combination of several elements that explains the general pattern:

- The consistently warm climates close to the equator permit year-round activity and allow greater niche differentiation. In contrast, some argue the marked

Climatic factors
1. Climate generally more benign and less variable in tropics, harsh and highly seasonal in higher latitudes
2. Under benign conditions shorter generation times promote higher speciation rates
3. Contrasting seasons at the higher latitudes limit close associations between species
4. Greater range of adaptations needed to exploit less benign conditions in higher latitudes; major groups originate in tropics and are still adapting to higher latitudes
Larger area
5. Greatest area of the globe found in the tropics, allows for larger populations and, therefore, fewer extinctions
6. Fragmentation of populations allows for higher speciation
Productivity
7. Higher net and year-round primary productivity in tropics allows for more elaborate food webs
8. Limited nutrient supply allows co-existence of many tropical plants in a small area
Constancy and history
9. Longer periods and larger areas undisturbed by climate change so that species accumulate and extinction rates are low
10. Longer period of co-evolution of species in tropics without major change in biotic or abiotic parameters
11. Moderate levels of disturbance may create a greater range of opportunities for colonists in the tropics, more distinct regeneration niche for trees in tropical forests
Complexity and stability
12. Complex plant communities provide more niche opportunities for animals in the tropics
13. High levels of competition, herbivory, and predation prevent any species from becoming dominant, and allow for stable species assemblages
14. Sexual reproduction is favoured to promote variation, especially to escape parasites
15. Greater diversity promotes greater diversity

Table 8.3 Possible factors contributing to the species richness of tropical forests.

seasonality of the temperate zones might sequence finer niche differentiation along a temporal gradient.

- Temperate communities cannot match the net primary productivity of the tropical ecosystems because of their winter. Less productivity constrains their food webs, or does not allow them to persist during the winter months and causes some animals to migrate.
- Certainly the seas of higher latitudes are at their most productive when nutrients and sunlight are abundant. Many tropical seas show only local maxima of productivity around the equator, where there is overturning of surface waters, bringing nutrients from depth.
- A shortage of the major nutrients is also a feature of tropical forests. There is a latitudinal gradient of soil fertility, and tropical soils have a low nutrient capital. Nutrient shortages are thought to prevent any single tree species becoming dominant in these biomes.

➔ *Section 7.2, Productivity and food chains; Section 8.5, The nutrient status of tropical soils*

This is supported by the contrasts between ecosystems within a climatic zone: nutrient-rich, highly productive estuarine and wetland communities have a few competitive, fast growing plant species; Mediterranean-type ecosystems are floral diversity hotspots growing on nutrient-poor soils.

High productivity does not guarantee more species, only more biomass.

➔ *Section 5.2, Tilman's hypothesis*

- Disturbance may prevent a few species dominating a community, especially in well-established ecosystems. To explain their species richness this would require rates of disturbance to be optimal in tropical regions, yet turnover rates for trees in temperate and tropical forest biomes do not differ. Some argue that the gaps created in tropical forests, with their high incident sunlight, provide a greater range of **regeneration niches** and opportunities for more plant species to colonize.

➔ *Section 6.4, Intermediate disturbance hypothesis*

- The tropical biomes occupy the largest land areas. Over large distances, populations are likely to differentiate into local ecotypes and eventually new species. Significantly, a tropical forest comprises different families and genera of tree, with no pure stands, whereas temperate forests have typically several species per genus, often in consistent stands. This suggests recent speciation, whilst the tropical forests indicate a longer evolutionary history, with little recent differentiation.

➔ *Box 6.4, Phylogenetic diversity*

- Large areas of deep oceans have undergone little abiotic change over millions of years. The **benthos** of a tropical sea has a low density of individuals but high species richness. These unchanging habitats harbour some of the oldest taxa. An unchanging environment causes fewer extinctions and, with time, greater specialization and niche differentiation. The tropics have had longer to accumulate species.

Global gradients of species richness

- The tropical forests were able to maintain large-scale refugia during the glaciations, where conditions remained largely unchanged. Today these represent 'hotspots' of biodiversity. Temperate ecosystems had to re-establish themselves after the ice retreated. Although many species have moved with the shifting climatic zones, others would need time to evolve the adaptations to the demanding conditions of the higher latitudes. Either way, higher latitudes have had a shorter period for niche differentiation and speciation.
- Temperate biomes encompass a wider range of abiotic conditions, favouring adaptable, more generalist species. Additionally, many species migrate seasonally to exploit resources elsewhere. Both may limit co-evolution between species and limit niche differentiation.

  ➔ *Section 2.2, Niche differentiation; Section 5.5, Co-evolution*
- Animal species richness in tropical forests and coral reefs rises with the structural complexity of these systems, and longer food chains are associated with greater three-dimensional complexity. The productivity and persistence of the tropics may thus foster positive feedback by increasing the ecological space.
  - Greater productivity allows for more structural complexity, and especially the vertical stratification seen in tropical forests and reefs. This allows for a wider variety of niches.
  - Adding more species creates opportunities: each becomes a resource or collection of niches available to other species. Thus, the diversity of tropical forests and reefs may follow from the diversity of resources represented by the biota itself.
  - The abiotic environment has been stable for longer in the tropics. Over time and with no major disturbance a complex community can develop.
  - Again, the lack of distinct seasons may allow most species associations to persist through the year.
- Extended periods of co-evolution may mean that parasitism, predation, and herbivory are more severe in tropical forests. A species consumed intensively is prevented from becoming dominant.
- Sexual reproduction may be favoured in the tropics because it promotes variation, possibly to avoid recognition by pathogens, parasites, and predators. Elsewhere, with fewer natural enemies, the benefits of sexual reproduction may not outweigh the costs; at higher latitudes, the frequency of asexual reproduction increases in some plant and animal groups.
- Thus species richness promotes species richness. Perhaps tropical ecosystems have simply had longer to build diversity.

The contrast between the latitudes probably follows from such positive feedbacks. As the community builds, new species and their interactions add to its complexity. Consumers constrain their prey, but also provide opportunities for others; limited nutrients constrain population growth.

With no species dominant, a more elaborate food web may be sustained by the year-round productivity of the tropics. It seems the variety of interactions between species, with connections of differing strengths, is key to the stability of such diverse ecosystems.

➔ *Section 6.4, Species–area relationships; Section 7.3, Stability and connectivity in ecosystems; Section 4.4, Population growth in a limited environment*

Make the connection

Pattern and scale in ecology

Notice how explanations of the global gradient in species richness draws on ecological principles considered earlier: principles explaining local phenomena are here used to explain these larger-scale patterns.

This means you can draw on smaller-scale examples to illustrate some of these ideas…and show how local processes relate to larger-scale processes. For example, does the evolution of ecotypes (Section 2.2), adapting to local conditions, explain the diversity of trees in the tropics?

➔ *Section 2.2, Ecotypes and ESUs*

8.3 REGIONAL AND LANDSCAPE ECOLOGY

Nutrient cycling or pollutant transfers are not confined to single ecosystems but follow atmospheric and regional processes. Landscape ecology recognizes that few ecosystems are discrete or isolated, and their position and history are fundamental to understanding their properties.

For example, acid rain is a problem in some regions of North America and Europe due to their position, climate, vegetation, and history. Most especially important are the regional air movements, the patterns of nutrient cycling, the effects of recent glaciations, the geology, and the nature of their soils. Elemental losses from acidified forests growing on cold and thin soils affect the freshwater and estuarine ecosystems in the northerly regions of these continents.

A landscape is a mosaic of terrestrial ecosystems which share a history, climate, and geology, and which exchange nutrients and species with each other. Many studies on nutrient transfer use watersheds as mesocosms: systems replicated within a landscape, with distinct boundaries defined by their drainage. These have helped identify the key processes in the acidification of these landscapes.

➔ *Section 7.1, Modelling ecosystem processes; Box 7.1, Mesocosms in systems ecology*

A landscape has three spatial elements (Box 8.2):

- the **patch**: a discrete ecosystem or habitat to which a species is adapted (for example, a pond);
- the **matrix**: the background ecosystem to which a species is not adapted (agricultural fields);
- **corridors**, which allow passage across the matrix, connecting patches (streams or hedgerows).
- Between these elements are more or less distinct boundaries or ecotones.

Box 8.2 Key technique

Quantifying landscapes

There are various ways of quantifying the grain and connectivity of a landscape but these measures have to change with the species being investigated, and the habitats and corridors it can use.

A suitable habitat represents a target for individuals dispersing from their home patch. Patches which are close, large, and with a long boundary are most likely to be reached by colonists (Figure 8.2). Populations in these habitats will have the greatest exchange of individuals and genes with the home patch. Corridors which allow a free flow of individuals increase the connectivity for that species.

These can be measured using detailed maps, aerial photography, or satellite imaging, although field work is invariably needed. This is because some corridors can be selective (filtering individuals with different dispersal abilities) and some patches are less easily colonized because of detail not recorded on maps or photographs. For example, a viral infection prevents amphibians from establishing in a particular pond, or seasonal temperatures block some corridors at certain times of the year.

The most direct measures are to simply count the numbers reaching a patch over a given interval and its population size. With large and easily observed species, ecologists can follow their movements directly, although various tracking technologies automate this for an increasingly wide range of animals. For plants and smaller animals, DNA and protein typing can help decide the parent population of colonists.

Because of their isolation, patches might be considered as habitat 'islands', and their species richness determined by their size and distance from a source community. Species–area relationships will then apply to different-sized patches with different degrees of isolation.

➔ *Section 6.4, Species-area and island biogeography*

Alternatively, we can use these metrics to compare landscapes in their entirety. Landscape physiognomy compares the arrangement of patches within a landscape and the distances separating them, especially important in species conservation: the exchanges between patches become important when local populations are liable to extinction or in the transmission of disease.

Landscape composition measures the proportion represented by each ecosystem, roughly equivalent to equitability in a measure of habitat diversity. The exchange of nutrients between them can indicate pollutant movements and loadings, and the susceptibility of local patches to regional pollution events.

- The extent to which a landscape is partitioned into distinct patches is its 'graininess'. The grain of a landscape changes according to the species of interest: an insect is likely to resolve a woodland into finer habitat patches than a bird.
- The degree of connectivity between patches is equally specific, according to the distances to be travelled between patches, the adaptations of the species, and their ease of movement along a corridor.

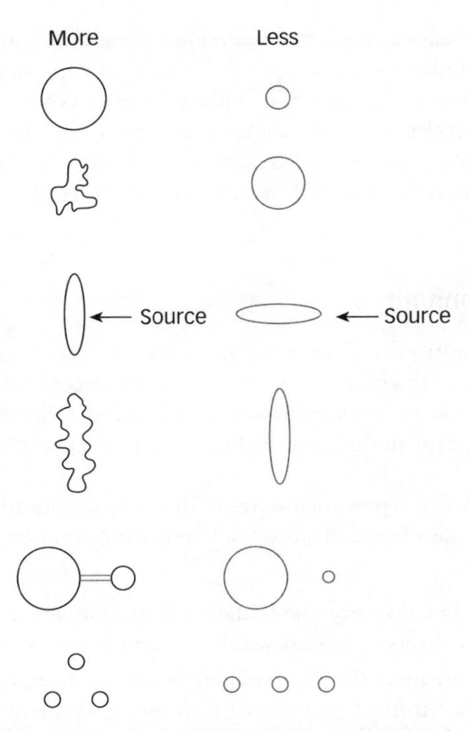

Figure 8.2 The connectivity within a landscape is determined by the size and shape of the habitats being studied. Individuals move freely between habitats if patch sizes are larger, have longer or more convoluted boundaries, represent a larger target, are connected by corridors, or have more neighbouring patches.

- Measures of habitat connectivity are needed to model the movement of individuals between populations, forming a **metapopulation**.
- Quantifying the grain and connectivity of a landscape can show how readily genes move between populations and why local populations may go extinct or undergo genetic drift.

  ➔ *Section 3.2, Genetic drift*

The landscape can also be the basis for modelling energetics and nutrient transfers.

- Adjacent ecosystems will exchange nutrients and species with each other. Some ecosystems are crucial sources or sinks of nutrients for other ecosystems. Within most landscapes, soils and aquatic sediments are the ultimate sinks for both nutrients and pollutants.
- Material can move within a landscape by earth movements (landslip, mass flow, etc.), by wind, and through transportation by water, both in surface and ground flow. The biota are often the principle means for long-distance movement of some nutrients.

  ➔ *Section 7.1, Modelling and systems ecology*

Regional and landscape ecology

A landscape is not fixed—its ecosystems develop through time, and may change from one patch type to another—such as a pond that gradually infills, or the sand dunes that become a pine forest. The history of a patch helps to explain the local ecology.

Often longer timescales and larger spatial scales are needed to understand these ecosystems: the drainage pattern, patch size, and connectivity of many northern temperate landscapes reflect the ice sheets that have covered them in the last 2 million years.

Transitional communities

Transitional communities form at the junctions between ecosystem types, where one abiotic factor gives way to another. The most obvious are between terrestrial and aquatic systems such as salt marshes, swamps, and estuaries. Some are very extensive and represent a landscape in their own right (e.g. deltas); others comprise just a few hectares.

Many such ecosystems repeat themselves in different regions of the world, and allow useful comparisons between latitudes. Comparing estuaries, for example, we find that:

- all tend to have a low diversity due to the harsh and variable conditions, but a wide phylogenetic diversity, at least within resident invertebrates;
- tropical estuaries are more diverse, probably because temperate estuaries are younger, and were reformed (or created) with the rapid rise in sea levels around 12000 years ago;
- all have low primary productivity. Their food web is based principally on their decomposer food chain and secondary productivity is high;
- where sea and river waters meet, water velocities slow and sediments are deposited. Here salinity varies according to the mix of the waters and the greatest variation in osmotic pressure occurs in these middle reaches (Figure 8.3);

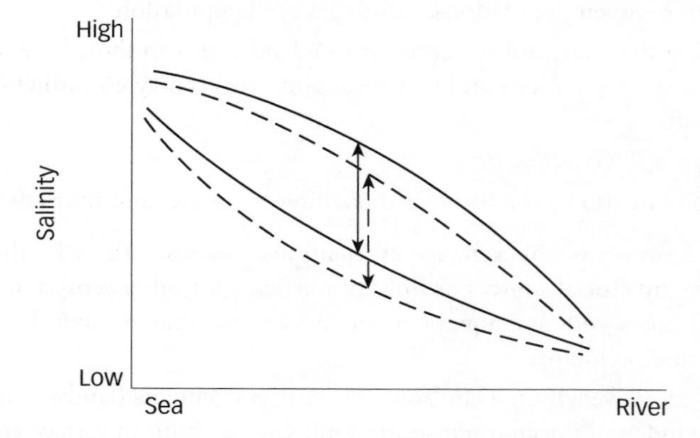

Figure 8.3 The variation in salinity along a tidal estuary. This range is largest in the middle reaches where fresh waters and seawaters mix. At the landward end, salinities are consistently low because tidal waters rarely reach this far. Similarly, at the estuary mouth marine waters dominate. These ranges change with season (broken and solid lines) and with the river's discharge.

- this creates demanding conditions for the biota and is thought to explain the pattern of low diversity among the invertebrates of the mudflats in their middle reaches.

However, estuaries differ:

- across the globe, by the large tidal ranges in the higher latitudes and the low salinity in the polar regions compared to the tropics;
- regionally by the size of the river's discharge;
- locally by the topography and geomorphology of the estuary mouth. This determines both the horizontal and vertical mixing of the fresh and salt waters.

In addition there are the temporal trends, through changing sea levels over the longer term and annually, according to the tides and seasonal river discharges.

Estuaries demonstrate both the effects of scale and local factors when comparing climatic regions.

Looking for extra marks?

A large range in key abiotic factors makes for a demanding habitat

Note (Figure 8.3) it is the range of salinities which estuarine organisms have to accommodate that make these demanding habitats, not salinity per se.

Make the connections here with the cost of accommodating that range: you are likely to score highly if you can describe the physiological costs associated with acclimation (Section 2.1), referring to osmoconformers and osmoregulators.

➔ *Section 2.1, Adaptation and acclimation*

8.4 LOCAL GRADIENTS AND THE STRUCTURE OF ECOSYSTEMS

Part of the complexity of an ecosystem results from the interaction of its major abiotic gradients: some are structured principally along a horizontal axis, such as the zonation across a tidal rocky shore or up an estuary. Riverine communities also change down their length, where the principal gradient is the gradient.

Revision tip

Example ecosystems

In the following sections we use three ecosystems to illustrate some general principles about structure and organization. You will have almost certainly studied other ecosystems in greater detail.

Consider using the perspective of scale and gradients to explore ecosystem organization and development as we do here. Attempt to apply these principles to your studies, and support them with relevant examples.

Stratification in temperate ecosystems

A range of ecosystems are structured by vertical gradients, often operating over small spatial scales.

Each stratum represents a distinct habitat, distinguished both by its position and its ecological processes. This is most obvious in soils, where the strata form **horizons**, zones of biological activity down the profile. Most terrestrial plant communities are stratified by gradients of light, moisture, and nutrients, as in forest communities.

Less obvious, and often only seasonally present, is the stratification of lentic water bodies, both freshwater and marine (Figure 8.4).

Stratification in temperate lakes

Lentic freshwater systems with clear waters allow light to penetrate to depth. Ponds and lakes then have upper waters which:

- are warmer than the lower layers;
- allow primary production by floating phytoplankton, as well as rooted and stratified **macrophytes.**

This temperature gradient and the clear conditions will persist as long as the sediments are not disturbed or until there is a pulse of material from outside the system.

- Water depth creates a gradient of temperature by which the lake can become stratified. The denser, colder waters sink and, in deep lakes, this stratification may be maintained throughout the year; some have many strata.
- A rapid change in temperature, known as the thermocline, can form within a few metres of the surface. When this develops in the summer, the upper and lower layers become isolated from each other and the supply of nutrients from sediments to the epilimnion ceases. This limits the productivity of this upper layer.
- While they persist these strata limit the exchange of water between different depths. Cold water stays close to the bottom and mixes little with the warmer upper waters.
- Consequently a gradient in available oxygen develops: anoxic conditions form in the waters next to the decomposing sediments and their community of anaerobic bacteria, fungi, and invertebrates. These conditions may persist for years in deep lakes.
- Rooted macrophytes can be important for securing nutrients from the sediments, supplying these to browsing species, and on their death, adding them back to the sediments. Animals feeding in the sediments may also add nutrients to the upper waters.
- Phosphorus is often limiting because it becomes immobilized in the sediments. With stratification, the productivity of the epilimnion depends on the phosphorus (and nitrogen) circulating within the living biota, or introduced from outside.

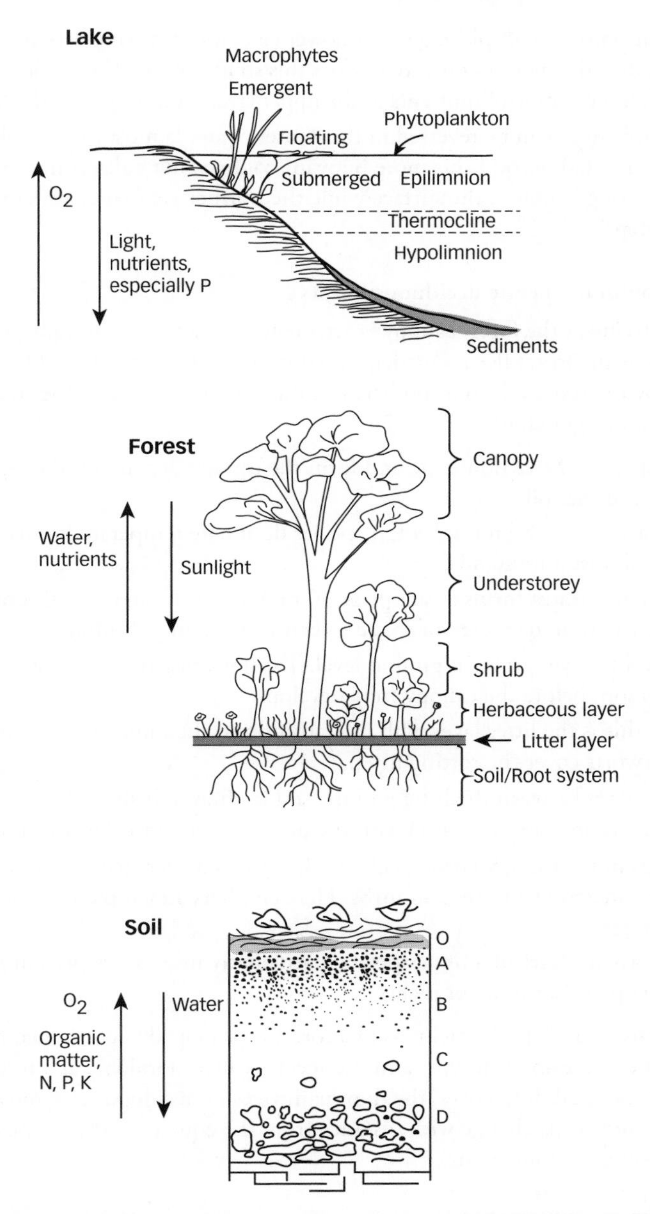

Figure 8.4 Stratification in three temperate ecosystems. A lake, a forest, and a soil have a vertical structure created by abiotic gradients. In all cases, the source of energy arrives at the top of the system (principally radiant energy in the forest and lake, organic matter in the soil) to create one gradient. Nutrients are held in the detritus and sediments towards the bottom of the first two ecosystems, whereas oxygen becomes limiting at depth in the soil and lake. These and other gradients create strata distinguished by their biological activity, dominant organisms, and associated metabolisms.

Soil horizons have different colours and texture because of the movement of organic matter and minerals and through their biological activity.

- Strong autumnal winds playing on shallower lakes forces mixing between the epilimnion and hypolimnion and destroys this stratification. This turnover replenishes the supply of nutrients to the upper layers and oxygen to the lower levels.
- The stratification can be reversed in the winter: water is most dense at 4°C, so as temperatures fall beyond this more buoyant colder water collects in the upper layer. Freezing begins at the surface while the warmer yet denser water remains fluid at depth.

Stratification in temperate deciduous forests

Sunlight arriving at the forest canopy is attenuated as its passes through successive leaves towards the forest floor. Nutrients held in the soil are transferred in water drawn up by the roots. The tension between these gradients creates the stratification seen in forests (Figure 8.4).

- Which species of tree dominate and form the canopy depend on the region, the climate, and the soils.
- In contrast to tropical forests, a few species dominate temperate forests and often occur in almost pure stands.
- A mature oak forest forms a canopy at 30 m, but this is rarely continuous, and beneath it is an understorey of shade-tolerant trees and a shrub layer.
- A herbaceous layer grows at ground level. The flowering of this layer occurs early in the season, before the canopy fully develops.
- Beneath this is the litter layer with its transition to the mineral soil. Mosses, ferns, and liverworts cover the rotting litter.
- The boundary between the litter and the soil is often indistinct. New gradients and a new stratification appears with the decomposer community of the soil.
- Openings in the canopy create glades with a different herbaceous community, often dominated by grasses and **forbs**. Here climbers attach themselves to the sides of trees.
- Discrete assemblages of animal species, particularly insects, are present in each layer, occupying a variety of niches.

The productivity and species richness of a forest is principally determined by regional gradients of climate and soil type, and the species which dominate as a result. Other than their geological differences, the dominant trees in deciduous and mixed temperate forests thus change with distance from the equator and their position relative to water-bearing winds.

Stratification in temperate soils

Plant and animal remains feed the decomposer community in the upper layers of a soil. Water arrives both from above and below, while oxygen has to pass into the soil to support aerobic decomposition. The soil–air interface thus creates a series of gradients, and generates a profile of declining colour and biological activity with soil depth (Figure 8.4).

- Except in waterlogged soils, aerobic decomposition dominates the litter layer (O horizon).
- Below this, in the dark brown of the A horizon, the products of decomposition first combine with the mineral components of the soil.
- Water percolating down the profile leaches the residues from A into the B horizon below. Most soluble and suspended matter is bound by the minerals here, although some will be lost with the groundwater. The B horizon is lighter because of the high proportion of minerals and reduced biological activity.
- The C horizon beneath is largely weathered bedrock with little organic matter and little biological activity.
- The invertebrate and microbial community changes with each strata and with the degree of oxygenation, especially in waterlogged soils.
- The strata are cut through by some biota, most especially plant roots, fungal hyphae, and deep-burrowing worms. The roots have a bacterial and fungal community associated with them (the **rhizosphere**), encouraged by root exudates, and this can discolour the lower layers.
- Water, oxygen, carbon, and nitrogen arrive at, or are fixed close to, the soil surface. Phosphorus and potassium have to be scavenged from below and are derived from the mineral and organic components of the soil.
- The organic matter of any soil is crucial for its water holding capacity and for much of its nutrient capital. The **cation-exchange capacity** (CEC) of a soil, its capacity to bind nutrients and pollutants, is largely determined by its **humus** and clay content.
- An absence, or an excess of water, can halt aerobic decomposition. In very cold and waterlogged soils decomposition can be slowed almost to a halt.
- Water is the main transport medium, moving soluble nutrients in a process termed **eluviation**. Rainfall is abundant in most temperate soils for most of the year so eluviation down the profile dominates.
- In warmer climates, where **evapotranspiration** exceeds precipitation, the predominant water movement is upwards. Salt and other soluble minerals may then crystallize at the soil surface.
- Minerals leached may be deposited in a lower horizon by **illuviation** (Figure 8.4).
- In wet and cold soils, where the pH is low, many minerals are solubilized but there is little illuviation. Nutrients are lost with the groundwater or held in the lower horizons of a waterlogged soil.

The physical properties of a soil (its texture and drainage) and its chemistry (its pH, buffering capacity, and CEC) determine its fertility. They result from the interaction between the overlying vegetation, the decomposer community and the parent material. Consequently they vary with the latitude and climate, geology, and history of a site and this explains one of the most important global gradients.

8.5 GLOBAL GRADIENTS IN SOIL FERTILITY

A mature soil is an integration of the abiotic and biotic factors operating on a landscape (Table 8.4). It summarizes the history of the site and its deeper layers often provide a record of its past. Indeed, we can reconstruct the duration and extent of the last ice ages from such deposits across the northern hemisphere.

- Many of the soils of the northern biomes are derived from the underlying igneous rocks that formed the ancient continent of Laurasia. These rocks are largely impermeable, slow to weather, and have been eroded by successive glaciations. The soils are thin and shallow with low CEC, and are thus susceptible to the effects of acid rain.

Soil type	*Biome/vegetation/ climate*	*Drainage*	*Texture/structure*	*Humus/soil pH*
Tundra peat/ gley	Bogs, thawed permafrost; swamps	Poorly drained	High proportion of poorly decomposed organic material; gleys when heavy clay	Mor
Podzol	High-latitude or high-altitude coniferous forests or heathland	Free, though periodic waterlogging at depth	Sandy	Mor, strongly acidic
Brown earth	Cool temperate deciduous woodland	Free	Loam, good crumb	Mull, going on slightly acidic
Rendzina	Temperate woodland or grassland on limestone	Highly leached	Thin	Mull
Chernozem (black soils)	Drier temperate grasslands	Evaporation exceeds annual rainfall	Good crumb	Alkaline upper layers because of calcium deposits
Latosol	Seasonally dry tropics, savanna grasslands	Often freely drained clays with low organic content; much of the silica leached out, leaving a high concentration of iron and aluminium	Shallow organic layer: brick-like laterites except where iron is leached away	Acidic, ancient, and highly weathered soils
Latosol	Humid tropics; red latosols dominate under tropical rain forests	Free, silica leached out, leaving a high concentration of iron and aluminium	Rapid decomposition so low organic content; poor, often clays that become brick-like when dry	Acidic, highly weathered

Table 8.4 A classification of the soil types associated with the major biomes from the pole to the tropics (omitting Mediterranean-type and desert biomes).

Mor soils are acidic, mull soils are alkaline: note that two types of temperate forest soil are shown according to the nature of the bedrock.

- The extensive plains of agricultural grasslands in the northern hemisphere were formed from the wind-blown deposits from the ice sheets of the Pleistocene. These are young and fertile soils, freely drained with a good texture.
- Most tropical forests and grasslands are situated on ancient and highly eroded plateaux. These old soils lack structure, nutrients, and, because of the rapidity of decomposition at these latitudes, organic matter. They are infertile and, when dry, form the brick-like **laterite** (Table 8.4).
- Without the protection of its vegetation, a tropical soil can quickly lose its small nutrient capital and its agricultural potential.

Globally, the result is a gradient of soil fertility between the temperate zones and the equator, reflecting their bedrocks, their history, and the current rates of nutrient cycling.

Yet the tropical forests growing on these thin and infertile soils are the single most important sink for atmospheric carbon on the Earth. Removing the forest removes the canopy, most of the biodiversity, and the decomposer community. With it goes the capacity of the ecosystem to capture carbon.

Make the connection

The larger picture

Take a moment to step back and look at the science you have been studying—some of the patterns we have used to structure this revision guide say something about ecology and evolution themselves—the timescales, the spatial scales, the interactions between different ecological compartments.

For example:

- Ecosystems and communities are structured and achieve a resilience in a time and across spatial scales which are, in large part, created by their biota: ecological space and time follow from the interactions between species. Towards the end of a succession, communities become stratified, more complex and more diverse, at a pace set increasingly by their biotic environment.
- Evolution has its pace set by rates of change at the molecular level (mutation, fixation), and by the autecology and geography of the species: their mobility, reproductive strategy, generation time, and so on.
- The fitness of individuals, the growth of populations, and the evolution of species all follow from an understanding of the species in its habitat. Even though we find it hard to define niche we can see that it captures the complementarity of a species with its ecological space.
- The interactions between individuals and species, and the effect of these on an organism's reproductive success, are responsible for the complexity of both the individual and the ecosystem. Through these interactions, and under these

continued

selective pressures, communities organize themselves and generate diversity. In this way life on Earth has evolved.

You may find it useful to read Chapter 1 (again) and ensure that you see all the connections pointed to in that chapter.

Check your understanding

Examination-type questions

1. What features of the species richness of tropical forests and the benthos of tropical deep oceans suggest a long evolutionary history? Can this explain their high diversity compared to that of temperate areas?
2. Devise a field survey you would conduct to measure the connectivity of ponds in a landscape, measured in terms of either:
   a. the fish in the ponds, or
   b. the waterfowl feeding in the ponds.

Give a clear justification for all of the parameters you would measure.

You'll find answers to these questions—plus additional exercises and multiple-choice questions—in the Online Resource Centre accompanying this revision guide. Go to http://www.oxfordtextbooks.co.uk/orc/thrive or scan this image:

Glossary

abiotic factors are the physical and chemical (non-biological) factors in its environment to which an organism has to adapt. cf **biotic factors.**

acclimation The phenotypic adaptation to a changed environment, by reversible physiological or morphological changes in an organism. cf **adaptation**.

adaptation A trait of an individual that enables it to survive or reproduce. Together, a collection of traits fits an organism to its environment. **Phenotypic adaptations** are physiological, behavioural, or developmental changes in an individual to accommodate adverse conditions. These changes are not passed on to its offspring and many are reversible. **Genotypic adaptations** are coded in the genes and therefore inherited.

admixture Gene flow between two populations each adapting to local conditions. This may be an important source of genetic novelty if they have been isolated for some time. cf **introgression**.

aerobic An environment that has sufficient available oxygen to support aerobic respiration, where oxygen is the final electron acceptor (oxidizing agent); aerobic metabolism uses oxygen, **anaerobic** metabolism does not.

age structure The proportion of individuals in different age classes of a population.

aggregate A group of closely related species or semi-species, usually applied to plants. *See also* **species**.

allele The various forms of a gene found at a particular position (*see* **locus**) on a chromosome, and which code for a particular trait.

allelopathy The inhibition of one organism by another using chemical means. Some plants may inhibit the growth of others (especially those of the same species) by the chemical nature of their litter or by special secretions.

allocthonous Originating outside an ecosystem.

allogenic and **autogenic succession** *See* **succession**.

allopatric speciation *See* **speciation**.

allozyme A structural variation of an enzyme, sometimes able to carry out the same function over a slightly different range of conditions.

altruism The donation of some advantage to others at the expense of the donor. In biology this translates into an increase in the recipient's fitness at a cost to the donor.

amino acid sequencing The chemical analysis of proteins determining the sequence of individual amino acids along the polypeptide chain. This can be used to measure the relatedness of different taxa by comparing the similarity between the sequences.

anaerobic (anoxic) An environment that is devoid of or has very low concentrations of oxygen; **aerobic** metabolism uses oxygen, anaerobic metabolism does not.

annual A species that completes its life cycle within a year.

antibody A protein produced by some animals in response to a foreign substance (**antigen**) as part of their immune system.

apomict A species that reproduces without fertilization. qv **microspecies**.

asexual reproduction Reproduction by the cloning of parental tissues. In asexual reproduction there is no swapping of genetic material between individuals.

assimilation efficiency The proportion of the energy assimilated into the tissues from the food consumed by an organism.

Glossary

assortative mating The tendency of individuals to mate with those with a similar genotype. This form of mate recognition reduces variation within a population and favours certain traits.

autecology The study of the relationship between a species and its environment; a species' natural history informed by its evolutionary history.

autotroph An organism able to obtain its energy either from sunlight (cf **photoautotroph**) or from chemicals (cf **chemoautotroph**).

authority The person who first describes and names a particular species.

basal metabolic rate The minimum rate of energy consumption for survival, usually measured in an organism at rest when there are no energy demands for acclimation or digestion.

Batesian mimicry Where a harmless species resembles a harmful or poisonous species to deter potential predators. qv **Mullerian mimicry**.

benthos, benthic communities Organisms living on the bed of an aquatic ecosystem. cf **pelagic communities**.

biennial A plant that lives for two years, with growth and establishment in the first followed by maturation, fruiting, and death in the second.

binomial system The system used to formally name species, using a generic name followed by the specific name (e.g. *Homo habilis*).

biochemical oxygen demand (BOD) A measure of the amount of organic matter in water.

biodiversity Often taken to simply be the number of species (cf **species richness**), but sometimes used to encompass biological variety at all levels, from the genetic diversity within a species to the total number of species on Earth.

biological species concept Species defined as a group of individuals able to breed together and produce viable fertile offspring. *See also* **species**.

biomagnification (Sometimes imprecisely equated with bioconcentration and bioaccumulation) The increase in the concentration of a pollutant in successive trophic levels; the concentration in a consumer divided by the concentration in its diet (or in plants, soil): a transfer coefficient.

biomass (standing crop) The mass of living or recently dead material, sometimes expressed as its energy equivalent.

biome The characteristic plant community associated with a particular latitude or biogeographical region, primarily determined by climate.

biosphere The part of the Earth that supports living organisms; the global ecosystem.

biotic factors are the biological factors in its environment, its interactions with other individuals or species, to which an organism has to adapt. cf **abiotic factors**.

Browser A consumer of shoots and leaves.

carrying capacity (*K*) The maximum number of individuals a habitat can support; a ceiling which cannot be exceeded over the long term. As this number is reached, environmental resistance, the effects of competition for resources, and the build-up of wastes, slows the growth of a population.

cation-exchange capacity (CEC) The total cations (positively charged atoms or molecules) that can be bound by a soil; a measure of its capacity to hold essential elements and buffer its acidity. CEC is high in organic and clay-rich soils, and is one indication of their fertility.

character displacement The divergence of a trait between populations over several generations, as a result of genetic change, and a consequence of adapting to different parts of an environmental gradient.

chemoautotroph A microorganism that derives its energy from oxidizing inorganic compounds and using carbon dioxide as its carbon source. Includes the nitrifying bacteria. cf **autotroph**, **photoautotroph**.

chemoheterotroph An organism which uses organic compounds both as a source of energy and of carbon.

chromosome The main site of the genetic material in living cells, composed of **DNA** and protein.

climax community The notional endpoint of a succession, when species turnover is minimal, and an equilibrium (or persistent) species assemblage is established. A **subclimax** community is one held at a stage prior to climax by a disturbance or some other factor. A **plagioclimax** community is one that has been deflected from its normal climax state as a result of human activity, and with a different plant community as a result.

cline A continuous change in a character within a series of contiguous populations along an abiotic gradient.

coefficient of selection (*s*) A measure of the intensity of selection against a genotype as the reduction in the proportion of gametes it adds to the gene pool, compared to a genotype under no selective pressure.

co-evolution The interdependent evolution of two or more species together.

cohort A group of individuals of the same age in a population.

commensalism An association between two species of benefit to one partner only, but of no detriment to the other.

community An assemblage of plants, animals, and microorganisms that persist together in the same habitat and interact with each other.

compartmentation The extent to which a community, or more specifically a food web, is divided into discrete units, with strong interactions between its members.

compensation The changes in the growth or physiology of a plant following losses to herbivores.

competition The fight for a limited resource between individuals of the same species (**intraspecific competition**) or between different species (**interspecific competition**).

competition coefficient A measure of the inhibition of the population growth of a species by each individual of a competing species.

competitive exclusion principle Two species cannot co-exist in the same niche at the same time in the same place. Of two species with identical resource requirements, one will eventually be ousted.

concentration factor The ratio of the pollutant concentration in an organism to that of its diet or the soil in which it grows.

congeneric Belonging to the same genus.

connectance The number of interactions between different species within a food web expressed as a proportion of the total number of possible interactions. This is the probability that any two species in the community interact with each other and a measure of the complexity of the ecosystem.

connectivity In landscape ecology, the extent to which habitat patches are connected to each other.

consumer A species that feeds on another species or organic matter; that is, herbivores, carnivores, parasites, and detritivores.

convergent evolution The independent evolution of two or more distinct species towards a similar adaptive solution under equivalent selective pressures, often resulting in similar morphologies or appearances. cf **parallel evolution**.

corridor In landscape ecology, a linear habitat that facilitates the movement of organisms across an inhospitable **matrix**, such as a roadside verge or a stream.

cost-benefit analysis Measures the balance between the costs and benefits of a trait. Natural selection should favour the optimum balance that maximizes **reproductive success**.

decomposer An organism that feeds on dead organic matter. Detritivores feed on **detritus**, fragmented organic matter. *See also* **saprotrophs.**

definitive host *See* **parasite.**

deme A local, interbreeding population. *See also* **species.**

denitrification The conversion of nitrate to nitrite and nitrite to molecular nitrogen by a range of anaerobic bacteria.

density-dependence Populations that grow in a density-dependent way increase their numbers in relation to the density of the population. Other organisms have density-independent growth.

detritus Fragmentary dead and decaying organic matter.

diploid The $2n$ chromosome state, having received one chromosome from each parent at fertilization. *See also* **haploid, polyploidy.**

directional selection A consistent change in the frequency of an allele within a population over several generations: its expression in the phenotype either increases or reduces fitness.

disruptive selection Occurs when an extreme phenotype enjoys higher fitness because of its difference from the population mean. cf **stabilizing selection.**

diversity Species diversity is usually measured as a combination of the **species richness** and the relative proportions of different species, or **equitability. Genetic diversity** refers to the totality of the variation in the **gene pool. Habitat diversity** is the number of niches (sometimes equated with the number of species), used to compare habitats by their structural complexity.

DNA Deoxyribose nucleic acid, the molecule that carries the genetic code. cf **RNA.** *See also* **gene, nucleotide.**

DNA sequencing A technique used to determine the sequence of bases along a piece of DNA.

domain The major taxonomic division of organisms, of which there are three: Archaea, Bacteria, and Eukaryota. *See also* **eukaryotes, prokaryotes.**

dominant species A species whose abundance or biomass are significant to the structure of the community. cf **keystone species.**

ecological gradient The change in some selective pressure, **biotic** or **abiotic**, over space or time, and along which niches may be differentiated.

ecological niche The totality of adaptations of a species to the biotic and abiotic factors in its environment; a species role in its community. *See also* **fundamental niche, germination niche, niche breadth, niche differentiation, niche overlap, niche packing, realized niche, regeneration niche, resource partitioning.**

ecological space The combination of all environmental gradients to which a species is adapted.

ecological time The temporal gradients to which a species is adapted.

ecology The science that studies the relationship between living things and their environment.

ecosystem A community of living organisms and its physical environment.

ecotone The zone between two adjacent communities.

ecotype A population closely adapted to localized conditions and with a distinct genotype from other members of its species.

effective population size The size of the breeding population at a particular time and location, taking into account the breeding behaviour of the species.

eluviation The translocation of dissolved (leaching) or suspended material by the movement of water through a soil profile. cf **illuviation.**

emergent property A characteristic of a population or community that could not be detected by looking at individuals, and which only emerges from seeing the assemblage operating together. One example is the shoaling behaviour of some fish.

endemic species Native to and only found in an area.

epidemiology The study of the spread of disease in a population.

eukaryotes Organisms whose cells have a discrete nucleus within a membrane. Eukaryotes include all algae, fungi, and all multicellular plants and animals. *See also* **domain**, **prokaryotes**.

eurytopic species can accommodate a wide range of conditions and have a broad geographical range. cf **stenotopic species**.

eusociality Animals living in highly integrated societies where non-reproductive individuals help raise offspring, such as ants, bees, and termites.

eutrophic ecosystems are nutrient-rich and have high primary **productivity**. cf **oligotrophic**.

eutrophication Changes in the structure of an ecological community as a result of nutrient enrichment, most often phosphate and nitrogen enrichment.

evapotranspiration The loss of water from terrestrial ecosystems, from soil and water bodies, and from plants via their transpiration stream.

evolution A directional change in the inherited characters of an organism. The process by which one species might arise from another (*see* **natural selection**). **Microevolution** refers to the changes which occur in the genetics of a population. **Macroevolution** refers to speciation and the phylogeny of species.

evolutionarily stable strategy (ESS) is one which cannot be beaten if most members of a population adopt it.

evolutionarily significant unit (ESU) A population considered to be on a separate evolutionary trajectory to other populations.

evolutionary species concept This recognizes that the genetic signature of a species changes through time and regards the species as a distinct lineage of adaptive changes (cf **evolutionarily significant unit, ESU**).

exaptation The presence of a trait, evolved under different selective pressures, which confers an advantage on a species colonizing a new ecological space.

exobiology The search for life beyond the Earth.

extant Living, as opposed to extinct.

exuviae Moulted shell, skins, or exoskeleton.

facilitation An effect where the activity or presence of one species allows other species to colonize a habitat, as part of the process of **succession**.

fecundity The capacity of an organism to produce viable offspring.

fitness In natural selection, **reproductive success**, measured as the proportion of the genome of the next generation represented by the individual's genotype.

flux rate The rate of turnover of an element through the biosphere.

food chain Consists of a series of compartments or **trophic levels**: primary producer, herbivore, carnivore. A **grazing food chain** is based on living plants as the primary producers and herbivores as the primary consumers. A **decomposer food chain** is based on decomposer organisms that feed on dead organic matter.

food web The trophic or feeding relations between different species of a community.

forb Broad-leaved herbaceous plant.

fossil record The entire catalogue of fossilized plants and animals that have been dated in geological time.

founder effect The genetic differentiation of a small population, unrepresentative of, and isolated from, its parent population.

frequency-dependent selection Occurs when the rarity of a form confers an advantage, for example in avoiding a predator.

frugivorous Fruit-eating.

fugitive species Species that maintain a population by colonizing ephemeral gaps. Usually applied to plant species that persist in habitats with a predictable frequency of disturbance cf **ruderal species**.

functional redundancy The duplication of roles within a community such that some species may be lost without any loss in ecosystem performance.

functional response A response of a predator to prey numbers in which it consumes more prey as they become available. cf **numerical response.**

fundamental niche The optimum ecological space, the set of conditions to which the species is best adapted. *See also* **ecological niche.**

gametes The sex cells, eggs (ova) or sperm; having a single set of chromosomes and therefore **haploid.**

gene A sequence of **nucleotides** on the **DNA** molecule that are inherited as a unit and which codes for a specific **RNA** molecule or a polypeptide for which that RNA molecule is responsible.

gene pool The totality of genetic material in an individual, or interbreeding population or species, at a particular time.

gene sequencing *See* **DNA sequencing.**

generalist species A species that can use a wide range of resources, typically undergoing ***r*-selection.** *See also* **specialist species.**

generation Average time between the birth (or equivalent) of the parents and the birth of their offspring.

generic name *See* **binomial system.**

genet An organism that grows from a fertilized egg and is therefore genetically distinct. cf **ramet.**

genetic drift The genetic differentiation of a small, isolated population through random changes and the sampling effect with each round of sexual reproduction.

genetic fixation The loss of all variation at one or more loci: fixation is complete when every individual in the population is **homozygous** for that locus.

genome The complete set of genes which define either an individual, a population, or a species, usually taken to be the **haploid** condition.

genomic analysis describes the gene sequences of species, populations, or individuals, and is used in systematics to determine the phylogeny of a group.

genotype The genetic code of an individual, or sometimes, rather confusingly, all individuals that share that code. cf **phenotype.**

genotypic adaptation *See* **adaptation.**

genotypic variation Variation between individuals that is wholly attributable to differences in their genetic code. Genetic diversity refers to the totality of the variation in the **gene pool.**

genus A category between species and family in which a number of closely related species are grouped.

geomorphology The study of landforms. *See also* **topography.**

germination niche The conditions necessary for a seedling to emerge and survive; more generally, the ecological space needed for juvenile survival. cf **regeneration niche.**

good gene hypothesis The suggestion that some genes confer traits which signal the fitness of the individual to potential mates. *See also* **sexual selection.**

grain In landscape ecology, the scale of resolution by which an organism resolves significant features of its habitat.

granivorous Seed eating.

greenhouse effect The warming of the Earth's atmosphere due to its capacity to absorb reflected long-wave (infrared) radiation while being largely transparent to incoming short-wave radiation.

gross primary production (GPP) The total amount of energy fixed by plants within a given area.

group selection occurs if traits within a group are favoured that benefit the group, rather than the individual, irrespective of genetic relatedness.

guild A group of organisms sharing a similar set of adaptations and carrying out a similar role or exploiting the same resource, within a habitat.

habitat The place where an organism lives; sometimes used in the general sense to refer to the type of place in which it lives.

haploid Having a single set of chromosomes; the $1n$ state, usually associated with a gamete (egg or sperm). *See also* **diploid**, **polyploidy**.

Hardy–Weinberg law The prediction that, in the absence of selection, genetic drift, mutation, or migration, random mating will cause allele frequencies to remain unchanged over generations.

herbivore An animal that feeds on plants.

heritability A measure of the total phenotypic variability in a trait attributable to genotypic variation in a population. Broad heritability is the proportion of phenotypic variation attributable to total variation in the genotype including any arising from interactions between genes; narrow heritability is the proportion of phenotypic variation that can be attributed to one or more individual genes (without regard for any interactions)—sometimes termed additive heritability.

heterotroph An organism that obtains its energy from other species; consumers.

heterozygote advantage follows when the expression of two different alleles at a locus increase the fitness of the phenotype.

heterozygous An individual that, for a particular character, has different genes on each of its paired chromosomes. cf **homozygous**.

homoiotherms maintain a constant body temperature. cf **poikilotherms**.

homozygous With the same allele for a trait on each chromosome. cf **heterozygous**.

humus The highly degraded organic matter within a soil.

hybrid A cross-bred individual derived from gametes from different populations, species, or genera.

hybrid breakdown A series of mechanisms that prevents hybrids from breeding.

hydrological cycle The movement of water from oceans onto land and back again.

igneous rocks Formed from the solidification of magma, the result of vulcanicity.

illuviation The precipitation and deposition of materials in the B horizon of a soil, moved by **eluviation**.

immunological methods Techniques that use antibodies to detect the similarity or difference between proteins and that rely on the specificity of the reaction between an antibody and its antigen.

inbreeding depression The reduced reproductive potential of a population in which individuals share much of the same genetic code. Conversely, **outbreeding depression** occurs when the genetic differences between individuals are too large to produce viable offspring.

inclusive fitness The total fitness of an individual resulting from its own reproductive effort and also that of genetically related individuals.

independent assortment The random allocation of **alleles** to the gametes at meiosis.

inhibition The capacity of one species to inhibit the colonization of another, especially in a successional sequence.

intermediate disturbance hypothesis The suggestion that maximum diversity occurs in communities subject to disturbance frequently enough to prevent competitive species becoming dominant but not so frequently as to depress colonization rates by immigrant species.

interspecific competition *See* **competition.**

intraspecific competition *See* **competition.**

introgression The influx of genes from one species or ecotype into the gene pool of another. cf **admixture.**

island-biogeography theory A model predicting a dynamic equilibrium of species number in a habitat, as a result of the opposing processes of colonization and extinction. This is allied to the **species–area relationship.**

iteroparity A life cycle with multiple reproductive events. cf **semelparity.**

joule The SI unit of energy; the amount of work needed to move 1 kg through 1 m.

K The carrying capacity, the maximum number of individuals for a population in a limited environment.

***K*-selection** Applied to an organism whose reproductive and **life history strategies** are primarily adapted to life in an unchanging and limited environment where there is intense competition for resources. cf ***r*-selection.**

key factor The source of mortality most likely responsible for the major fluctuations in total population size.

keystone species A species whose presence or activity determines community structure yet their numbers are usually small. cf **dominant species.**

kin selection The favouring, or disfavouring, of genes shared by closely related individuals.

kinetic energy Energy of a body due to its motion.

Kingdom Five broad taxonomic categories, subsets of the two Superkingdoms: Prokaryota and Eukaryota.

latitude Lines of latitude run parallel to the equator: lower latitudes are close to the equator, higher latitudes closer to the poles. Lines of **longitude** radiate from the poles, counted from the Greenwich meridian.

landscape ecology The study of ecological processes across several ecosystems united by a shared landform, climate, and regime of disturbance. Adjacent landscapes under an equivalent climate or topographical area may be collected together as a region.

laterite A hard red-brown clay, devoid of many soluble minerals and nutrients, and which forms in tropical areas dominated by ancient soils.

leaching The loss of material when it is washed out of a soil.

lentic A freshwater system with slow-moving or stationary water (ponds and lakes). cf **lotic.**

life history strategy The allocation of time and resources that an organism makes between different stages of its life cycle so as to maximize its reproductive potential. *See also* **reproductive strategy.**

life table A summary of the rates of mortality and survivorship of different age groups in a population.

loam A well-balanced mixture of sand, silt, and clay with abundant organic matter producing a well-drained soil with a good **soil crumb** structure.

locus The site of an individual gene on a chromosome. *See also* **allele**.

lotic Fast-moving freshwater system (rivers and streams). cf **lentic**.

macrophyte An aquatic plant that is either rooted or attached to a submerged substrate and with a stem that holds the leaves close to or above the water surface.

matrix In landscape ecology the dominant component of the landscape; for example, agricultural fields. Often an inhospitable space that the organism of interest has to cross to reach habitat patches.

meiosis The division of the paired chromosomes necessary to form a gamete. cf **mitosis**.

mesocosms Division of a natural ecosystem (such as a lake or watershed) into a series of smaller units to allow for experimentation. *See also* **microcosms**.

metapopulation A population of populations. A series of populations that may swap individuals with each other, but are usually divided into discrete patches.

microcosms Small, artificial ecosystems that attempt to mimic some features under controlled conditions, to allow for manipulation and replication. *See also* **mesocosms**.

microspecies A plant able to produce seed without pollination (fertilization) and which is consequently genetically isolated.

Milankovitch cycles generate long-term changes in the Earth's climate, and derive from changes in the shape of the Earth's orbit, the tilt of its axis, and the precession of its equinoxes.

mitochondria Organelles, the site of cellular respiration in eukaryotes.

mitosis Division of the nucleus but not the chromosomes, as a preclude to cell division. cf **meiosis**.

Monera The taxonomic Kingdom comprising prokaryotic single-celled organisms which lack a true nucleus and membrane-bound organelles.

mor soil type Mor soils are acidic forest soils, typical of cold, wet soils under coniferous forest where decomposition is slow. cf **mull soil type**.

morphological species (also known as **typological species**) Described by their morphological characteristics alone and the basis of most multicellular classification. *See also* **species**.

mortality rate (*m*) The death rate.

mull soil type Mull soils have their organic matter incorporated into the mineral soil, are relatively dry, and neutral to alkaline. cf **mor soil type**.

Mullerian mimicry Shared patterns and colouration adopted by potential prey that signal danger to predators.

mutation A change in the genetic code that may be inherited.

mutualism A form of symbiotic relationship (*see* **symbiosis**) between two species to their mutual benefit.

mycorrhiza A symbiotic association (*see* **symbiosis**) between plant and fungus, where the plant root acts as a host to fungal threads.

natality rate (*b*) The birth rate.

natural selection The reproductive success of different individuals under the constraints placed upon them by their environment. Less-fit individuals fail to reproduce and their genes are lost from the **gene pool**. The persistence of particular adaptive traits through the generations produces the inherited change that may produce new species.

neodarwinism A re-statement of the principle of evolution by natural selection, incorporating our understanding of inheritance and population genetics.

net primary production (NPP) The amount of energy fixed within plant tissues after metabolic and photosynthetic costs have been met.

net reproductive rate *See* **reproductive rate (R_o)**.

neutral theory of evolution The suggestion that populations can maintain genetic differences originating from random (non-selective) processes.

neutral variation Genetic variation within a population or species that confers no selective advantage.

niche *See* **ecological niche.**

niche breadth The range of an environmental gradient or resource over which a species can survive and reproduce. *See also* **ecological niche, tolerance limits.**

niche differentiation The partitioning of an ecological space between two or more species, each exploiting a different characteristic of the resource. *See also* **ecological niche.**

niche overlap That part of a resource gradient that two or more species occupy (for which they may or may not be competing). *See also* **ecological niche.**

niche packing The number of species occupying a resource gradient in the same habitat at the same time. *See also* **ecological niche.**

nitrification A process carried out by a range of soil microbes by which ammonia is oxidized to nitrite, and nitrite is oxidized to nitrate.

nucleotide The basic unit of **DNA** and **RNA** comprising a phosphate group, a sugar group, and a nitrogenous base.

nucleus In eukaryotic cells, the organelle in which chromosomes are held.

numerical response The increase in numbers of a predator as a response to the increased abundance of its prey. cf **functional response.**

oligotrophic Nutrient-poor; oligotrophic ecosystems have low primary **productivity**. cf **eutrophic.**

omnivore An animal that feeds on plants and animals.

optimality theory The suggestion that an organism will adopt a strategy, behaviour, or adaptation that will maximize its return on its efforts or use of resources, measured by its **fitness**. Often applied to foraging or predation.

osmoconformer An organism that does not regulate the osmotic concentration of its internal fluids and so changes with external osmotic concentrations. cf **osmoregulator.**

osmoregulator An organism that maintains its internal osmotic concentration. cf **osmoconformer.**

parallel evolution The independent evolution of the same trait from a shared ancestral trait, in two or more species long separated geographically. cf **convergent evolution.**

parapatric speciation *See* **speciation.**

parasite An organism that is metabolically dependent on another, at the expense of the host. A **definitive host** is one in which the parasite matures (and may reproduce sexually); an intermediate host serves as a **vector** in transmitting immature stages between definitive hosts.

parasitoid Insect **parasites** (primarily Hymenoptera and Diptera) that lay an egg within a host insect and which eventually leads to the host's death. Often seen as a specialized form of parasitism and predation.

patch Element(s) of an organism's landscape that it can utilize, such as a pond or woodland.

pathogen An organism or virus that can cause disease.

pelagic communities Organisms living in the open water of an aquatic ecosystem. cf **benthos, benthic communities.**

perennial A plant that lives for more than two years.

permafrost The soil layer found at depth beneath tundra vegetation that remains frozen throughout the year.

pest A species whose presence causes a nuisance and results in some economic cost.

phenology The timing of biological events, such as flowering or breeding, in relation to ecological time, especially seasons and climate.

phenotype The characteristics of an individual, the product of the expression of its genetic code and its interaction with the organism's environment. cf **genotype**.

phenotypic adaptations *See* **adaptation**.

phenotypic plasticity The capacity of an organism to change physiology or morphology to accommodate the prevailing environmental conditions, changes which are not inherited; often referring to significant variations in form between populations from different habitats which are not reversible.

phenotypic variation Variation between individuals that is wholly attributable to physiological or developmental (i.e. non-genetic) differences.

photoautotroph An organism that obtains its energy from sunlight through the process of photosynthesis. cf **autotroph**, **chemoautotroph**.

photoheterotroph Bacteria which use radiant energy and organic sources of carbon in their energy fixation.

phylogeny The evolutionary relationships and history of a **taxon**.

poikilotherms allow their body temperature to vary with ambient temperatures. cf **homoiotherms**.

pioneer species Those first to colonize a new or disturbed site.

plagioclimax community *See* **climax community**.

polymorphism Variation within a species where individuals may take different morphologies.

polyploidy The state of having more than two sets of chromosomes. May occur when more than one of each chromosome is passed to the offspring. For example tetraploids are $4n$, having received $2n$ from each parent. *See also* **diploid**, **haploid**.

population Individuals of the same species living in a defined area at a defined time.

population density The number of individuals per unit area (or volume).

post-zygotic barrier Some failure in the subsequent development of the fertilized egg or the individual which makes it non-viable, unable to survive or reproduce.

potential energy Energy that is stored (e.g. chemically as starch and oils in plants, or fat and glycogen in animals).

predator An animal that kills another animal to feed.

pre-zygotic barriers A range of mechanisms that prevent the formation of a hybrid zygote from gametes from two different species or varieties.

primary producer An **autotroph**. *See also* **secondary producer**.

primary production The synthesis of complex organic molecules from simple inorganic ones by **autotrophs**, using either sunlight (**photoautotrophs**) or chemical energy (**chemoautotrophs**). cf **secondary production**.

primary succession *See* **succession**.

prokaryotes Single-celled organisms without a well-defined nucleus; the Bacteria and the Archaea. *See also* **domain**, **eukaryotes**.

Protista The taxonomic Kingdom that includes protozoa, primitive algae, and fungi.

production The biomass or energy fixed in the tissues by an individual or population, or per unit area or volume for a trophic level or ecosystem. **Gross production** is that measured before respiration; **net production** is after respiratory and excretory losses.

production efficiency A measure of efficiency by which an organism converts the energy from its food into tissues (the ratio of the amount of energy fixed in its tissues to the amount consumed).

productivity The energy fixed in the biomass of an individual, population, species, or trophic level.

proximate explanations explain how an adaptation is achieved.

qualitative traits occur as discrete phenotypes, often presence or absence.

quantitative traits show continuous variation between individuals, resulting from both genetic and environmental variation.

***r* (*r_m*)** The intrinsic rate of change in a population per individual; the instantaneous rate of change per individual in unit time. r_m is the maximum possible rate of population growth per individual under ideal conditions.

***r*-selection** Organisms that are *r*-selected are primarily adapted to life in changeable, short-lived habitats, where rapid population growth is favoured. cf ***K*-selection**.

ramet An individual that has arisen by asexual reproduction. cf **genet**.

realized niche The ecological space to which a species is confined by its interactions with other species. *See also* **ecological niche**.

recombination The swapping of fragments of genetic code between chromosomes during meiosis (sometimes called crossing over) and which may therefore produce a new **genotype**.

regeneration niche The conditions created by a disturbance and the loss of established species; these may represent an ecological space that other species are able to colonize. cf **germination niche**.

replacement probability In community ecology, the chance that an individual will be replaced by another of the same species within a given time. A dominant species has a high replacement probability.

reproductive rate (*R_o*) The average number of offspring per individual in a given time. The net reproductive rate (R_N) is the average number alive per individual in the next time interval, allowing for births, deaths, and survivors. In epidemiology the net reproductive rate (R_I) is the number of individuals infected by each infected individual.

reproductive strategy The timing and allocation of resources to reproduction during the life cycle of a species, refined by natural selection to maximize its **reproductive success**. *See also* **life history strategy**.

reproductive success The number of offspring produced by a female up to the present. *See also* **fitness**.

reproductive value The number of offspring produced by a female in the future; **total reproductive success** is the current plus the future reproductive output from a female.

residence time *See* **turnover time**.

resource partitioning The division of a limited resource between species, reducing any competitive interaction. *See also* **ecological niche**.

resource spectrum The range of a resource that is available to organisms in a habitat. Different species may use different parts of the spectrum, such as food items of different sizes.

rhizosphere The soil habitat adjacent to and influenced by the root system of a plant.

ribosome Site of protein synthesis in the cell.

riffle The shallow waters flowing over beds of gravel in a stream.

RNA Ribonucleic acid. A polynucleotide molecule that exists in several forms in the cell and carries out various functions, especially in protein synthesis. cf **DNA**. *See also* **gene**, **nucleotide**.

ruderal Plant species characteristic of disturbed and temporary habitats.

rules of assembly The suggestion that there are certain ways in which a community can be configured, or certain combinations of species, particularly regarding the structure of food webs.

ruminant Mammalian cud-chewing herbivores with a chambered stomach in which a culture of cellulolytic bacteria is maintained.

saprotrophs Organisms that derive their nutrients from dead and decaying organisms.

secondary consumer An organism that is dependent on primary consumers (herbivores) for its energy needs.

secondary plant metabolites Compounds produced by plants that have a defensive role against herbivores.

secondary producer An organism that derives its energy from a **primary producer**.

secondary production The energy fixed in the tissues of heterotrophs. cf **primary production**.

secondary succession *See* **succession**.

segregation The separation of homologous chromosomes at **meiosis** into separate cells.

selection differential (*S*) The mean difference between a selected trait and its mean value in the larger population.

selection response (*R*) The mean difference in a trait between the offspring of a selected group and the offspring from the larger population.

semelparity A life cycle with a single reproductive event. cf **iteroparity**.

semi-species Incipient species, a race showing weak reproductive isolation from the rest of a population. *See also* **species**.

sere The sequence of plant communities that lead to a climax plant community.

sexual dimorphism Phenotypic differences between the sexes that have adaptive significance.

sexual reproduction Reproduction that requires the fusion of two **haploid** gametes to form a **diploid** cell in a process of fertilization.

sexual selection The selection of individuals according to particular traits by a potential mate, when inherited traits increase or reduce the chances of mating, in competition with other individuals of the same sex.

sink In **systems ecology**, a reservoir or long-term repository of a nutrient or pollutant.

soil crumb The crumb structure of a soil is its capacity to form small cohering lumps. Soils with a good crumb will not waterlog but will have a high water-holding capacity.

soil horizon A distinct layer (usually distinguished by colour and texture) in a soil profile that demarcates a zone of biological or chemical activity.

specialist species Sometimes referring specifically to diet, but also to *K*-selected species closely adapted to their habitat and a particular part of a resource spectrum. Invariably highly competitive species. *See also* **generalist species**.

speciation The formation of new species as a result of evolution by natural selection. Speciation can be **allopatric** through geographical isolation, or **sympatric** when genetic isolation occurs between populations with overlapping distributions. **Parapatric** speciation is where adjacent populations diverge, say along some environmental gradient or by exploiting different resources.

species A collection of individuals able to breed with each other and produce viable (fertile) offspring. Note this functional definition is used to demarcate the fundamental unit in taxonomy and is the basis of the **biological species concept**. *See also* **deme**, **semi-species**, **syngameon**.

species–area relationship A highly consistent pattern of increase in species number (or **species richness**) with larger area or sample size.

species equitability The proportion of individuals in each species, which is used in combination with *S* in some indices to measure **diversity**.

species richness, species diversity The number of species in a habitat or S.

stability The capacity of a system to resist change (**inertial stability**) or to return to its original position or original rate of change following a disturbance (**elasticity** or **adjustment stability**).

stabilizing selection The favouring of the common form by natural selection, working against extreme phenotypes. cf **disruptive selection**.

stable age distribution This occurs when age-specific birth rates and death rates become fixed within a population. The population is then growing exponentially.

stable isotope analysis A technique, using radioactive isotopes that do not decay rapidly, to trace the trophic transfers through an ecosystem.

stationary age distribution occurs when overall birth rates and death rates are balanced so there is no population growth.

stenotopic species are adapted to a narrow range of environmental conditions and are confined to a narrow geographical range. cf **eurytopic species.**

strategy A programme for allocating effort or resources to maximize fitness, refined by natural selection.

stress-tolerators A category of plants that have a life history strategy adapted to life in hostile conditions.

subclimax community *See* **climax community.**

succession The progressive change in the species composition of a community with time. A **primary succession** is where there has been no previous life on a site; a **secondary succession** re-establishes a community where some nutrients or organic matter remain from its previous occupants. An **autogenic succession** is primarily governed by species interactions; an **allogenic succession** is governed principally by abiotic factors.

symbiosis The association of two species living together. A variety of associations are possible. *See also* **mutualism.**

sympatric speciation *See* **speciation.**

syngameon A collection of sympatric **semi-species.** *See also* **species.**

systematics The organization of species into a hierarchical classification that reflects their phylogeny.

systems ecology analyses ecosystems by their processes or functions. A system consists of two or more interacting elements that persist together within some discrete boundary.

taxon (*pl.* taxa) A taxonomic grouping, at any level in the classification hierarchy.

taxonomic distance The number of taxonomic steps separating two individuals, or the average for an entire sample of a community.

taxonomy The description, naming, and classification of organisms.

threshold density (S_T) In epidemiological models, the density (or number) of susceptible individuals needed for the disease to persist, when the rate of transmission is one per infected individual.

tolerance (i) The ability of a species to survive adverse or toxic conditions; (ii) in a succession, the capacity of many late successional plants to persist in poor growing conditions, typically low nutrient levels.

tolerance limits define the **niche breadth** of a species.

topography The study of surface features of an area. *See also* **geomorphology.**

trade-off The compromises made in adapting to two conflicting selective pressures.

transcription The creation of a molecule of messenger **RNA** using part of a **DNA** molecule as a template. *See also* **translation.**

transcriptome All the genes that are expressed in the phenotype from a particular **genome.**

transit time *See* **turnover time.**

translation The conversion of the code of the messenger **RNA** to create a chain of amino acids.

translocation (*genetics*) The movement of part of a chromosome to a different part of the same chromosome or to become attached to another chromosome, through an error in the natural copying and replication processes of the cell.

transpiration The evaporation of water vapour from the aerial parts of a plant.

trophic level A compartment or category of feeding relations, defined by its position along a food chain. A trophic pyramid refers to stacked trophic levels, with each level proportional in size to its amount of energy, biomass, or number of individuals.

turnover time (also known as residence time or transit time) The rate of energy or nutrient transfer; the time required to replace the standing crop.

type specimen The reference specimen(s) for a **typological** or **morphological species**, against the description of which other specimens are compared. Also known as the **haplotype**.

typological species (also known as **morphological species**) Described by their morphological characteristics alone and are the basis of most multicellular classification. *See also* **species**.

ultimate explanations explain why an adaptation is advantageous.

variation Differences between individuals of the same species. **Phenotypic variation** is acquired, through growth, development, or physiological change, but is not passed on in the genes. **Genotypic variation** are differences coded in the genes, which may therefore be passed on to the next generation. **Discontinuous (qualitative) variation** occurs when a trait is either present or absent; **continuous (quantitative) variation** occurs when a trait can show a range of expressions in the phenotype.

variety A distinct form within a species that occurs naturally.

vector An organism that carries a **parasite** from one host to another, also known as an **intermediate host**.

vegetative propagation Producing new individuals without sexual reproduction; in horticulture, producing individuals by promoting asexual reproduction.

Index

D

Index

N

O

Index